Breeding and Genetic Engineering

The Biology and Biotechnology Research

Breeding and Genetic Engineering

The Biology and Biotechnology Research

Published by iConcept Press

Breeding and Genetic Engineering – The Biology and Biotechnology Research

Publisher: iConcept Press Ltd.

ISBN: 978-1-922227-355

Printed in the United States of America

𝒾Concept
Press Ltd.

www.iconceptpress.com

Contents

Jorge A. Marquez-Escalante (*Centro de Investigación en Alimentación y Desarrollo, AC., Hermosillo, Sonora, Mexico*), Elizabeth Carvajal Millan (*Centro de Investigación en Alimentación y Desarrollo, AC., Hermosillo, Sonora, Mexico*), Yolanda L. López-Franco (*Centro de Investigación en Alimentación y Desarrollo, AC., Hermosillo, Sonora, Mexico*), Jaime Lizardi-Mendoza (*Centro de Investigación en Alimentación y Desarrollo, AC., Hermosillo, Sonora, Mexico*), Elisa Valenzuela-Soto (*Centro de Investigación en Alimentación y Desarrollo, AC., Hermosillo, Sonora, Mexico*), Agustín Rascón-Chu (*Centro de Investigación en Alimentación y Desarrollo, AC., Hermosillo, Sonora, Mexico*) and Craig Faulds (*Aix Marseille Université, Marseille, France*)

Júlio César Nepomuceno (*Universidade Federal de Uberlândia, Brazil / Centro Universitário de Patos de Minas, Brazil*)

Raquel Soares Casaes Nunes (*Universidade Federal do Rio de, Rio de Janeiro, Brazil*), Eduardo Mere Del Aguila (*Universidade Federal do Rio de, Rio de Janeiro, Brazil*), Vania Paschoalin (*Universidade Federal do Rio de, Rio de Janeiro, Brazil*) and Joab Trajano Silva (*Universidade Federal do Rio de, Rio de Janeiro, Brazil*)

Yeasmin Jolly (*Bangladesh Atomic Energy Commission, Bangladesh*), Roksana Huque (*Bangladesh Atomic Energy Commission, Bangladesh*), Ashraful Islam (*Bangladesh Atomic Energy Commission, Bangladesh*), Safiur Rahman (*Bangladesh Atomic Energy Commission, Bangladesh*), Shirin Akter (*Bangladesh Atomic Energy Commission, Bangladesh*), Jamiul Kabir (*Bangladesh Atomic Energy Commission, Bangladesh*), Kamruzzaman Munshi (*Bangladesh Atomic Energy Commission, Bangladesh*), Mahfuza Islam (*Bangladesh Atomic Energy Commission, Bangladesh*), Afifa Khatun (*Bangladesh Atomic Energy Commission, Bangladesh*) and Arzina Hossain (*Bangladesh Atomic Energy Commission, Bangladesh*)

Kadur N. Guruprasad (*Devi Ahilya University, Indore, India*), M. B. Shine (*Devi Ahilya University, Indore, India*) and Juhie Joshi (*Devi Ahilya University, Indore, India*)

Preface

Since the first genetically modified crop plant was produced in 1982, the discovery and improvement of plants and crops species based on breeding and genetic engineering has never been stopped. This book focuses on various aspects of plants genetics and plant breeding, molecular biology crop reproduction, soils and plant nutrition, and environmental related issues. It does not only highlights the current research issues in the area of biology but also the development issues of plants and crops in biotechnology. This book is recommended for experts in the field of botany, agriculture, and genetics.

Chapter 1 studies the swelling and microstructure of Tacupeto F2001 (a spring wheat variety in Northern Mexico) arabinoxylans gels. By using immersion in liquid nitrogen (fast congelation) before lyophilization, Tacupeto F2001 arabinoxylans gels present cells average inner dimensions lower than those reported by using slow congelation.

Chapter 2 reviews studies in humans and animals in order to evaluate the use of lapachol and its derivatives as a therapeutic intervention in cancer patients.

Chapter 3 study the taxonomy and phylogeny of Brazilian cultivars of *Colocasia esculenta*. Analysis of the chloroplast genome sequences such as *rbcL* and *pbsA-trnH* can be a valuable tool in establishing the phylogenetic analysis and variability of taro cultivars grown in Brazil.

Chapter 4 entails the presence of toxic elements (Cr, Co, Ni, Cd, Pb, As, Cu, Zn, Mn) in Rice of Bangladesh, which is the staple food of the country. In this chapter the possible source of toxic element which can increase the concentration in rice like water from the rice field, soil where rice plant grown were also analysed and possible potential risk of those elements to human health was also calculated to give a picture of the present status of rice in Bangladesh.

Chapter 5 discusses the effects of magnetic field on crop plants. Magnetic field may provide a feasible non-chemical solution in agriculture, meanwhile may offer advantages to protect environment and safety for the applicator.

Editing and publishing a book is never an easy task. Each chapter in this book has gone through a peer review, a selection and an editing process so as to guarantee its quality. Without the supports and contributions of the authors and reviewers, this book can never be able to complete. We would like to thank all of the authors in this book and all of the reviewers who participated in the reviewing process: Sharmila Chattopadhyay, Jinping Chen, Raphael Richard Ciuman, José G. Crespo, Nune Darbinian Sakkissian, VJ Galani, A.A. Haleem Khan, Parvez I Haris, Jose Antonio Hernandez, Erebi P. Ikeh-Tawari, Nafees A. Khan, Chang-Seok Ki, Mei-Chin Lai, Ramona Lall, Lijia Li, Jung-Yaw Lin, Andrew Lowe, Neeti Sanan Mishra, Sadegh Mohajer, Guogui Ning, William Parker, Anna Preová, Shahedur Rahman, Michael Shintaku, Kenta Shirasawa, Haitham Sobhy, Florencio M. Ubeira, Daisei Ueno, Aimin Zhang and Haili Zhang. We hope that you, the reader, will find this book interesting and useful. Any advices please feel free and are always welcome to tell us.

iConcept Press Editorial Office
July 2016

Gels of Water Extractable Arabinoxylans from A Bread Wheat Variety: Swelling and Microstructure

Jorge A. Marquez-Escalante[1], Elizabeth Carvajal-Millan[1],
Yolanda L. López-Franco[1], Jaime Lizardi-Mendoza[1],
Elisa Valenzuela-Soto[1], Agustín Rascón-Chu[1] and Craig Faulds[2]

1 Introduction

The development of high-quality wheat (*Triticum aestivum* L.) cultivars depends on a thorough understanding of the constituents of grain as the biochemical constituents of wheat grain largely determine its end-use quality. One such important constituent in grains is a group of nonstarch polysaccharides referred to collectively as pentosans. More specifically, a major member of the pentosan family, albeit a minor constituent in grain overall, is a polymer known as arabinoxylans (AX). Tacupeto F2001 is a spring wheat variety developed by The International Maize and Wheat Improvement Center (CIMMYT) in Mexico, and provided to National Institute for Investigation in Forestry, Agriculture and Animal Production (INIFAP) for testing and release. INIFAP released this wheat variety for cultivation in Northwestern Mexico. Tacupeto F2001 is a bread wheat variety presenting resistance to leaf rust. Previous studies have examined the quality parameters of Tacupeto F2001 wheat. However, to our knowledge, there are no reports about Tacupeto F2001 AX gels swelling. AX are

[1] Research Center for Food and Development, Centro de Investigación en Alimentación y Desarrollo, AC., Hermosillo, Sonora 83000, Mexico

[2] Laboratoire de Biotechnologie des Champignons Filamenteux, Polytech Marseille, Marseille, France

non-starch polysaccharides from the cell walls of cereal endosperm constituted by a linear β-(1→4)-xylopyranose backbone and α-L-arabinofuranose residues as side chains on O3 and O2 and O3 (Izydorczyk & Biliaderis, 1995). AX can present some of the arabinose residues ester-linked on (O)-5 to ferulic acid (FA) (3-methoxy, 4 hydroxy cinnamic acid). Dehydrodimers of ferulic acid (di-FA) structures may serve to cross-link cell-wall polymers and contribute to the mesh-like network of the cell wall (Niño-Medina et al., 2010). AX can gel by covalent cross-linking involving FA oxidation by some chemical or enzymatic (laccase/O_2 and peroxidase/H_2O_2 system) free radicals-generating agents (Geissman & Neukom, 1973; Hoseney & Faubion, 1981; Figueroa-Espinoza & Rouau, 1998). Five main di-FA (5-5', 8-5' benzo, 8-O-4', 8-5' and 8-8' di-FA) are identified in gelled AX, the 8-5' and 8-O-4' forms being generally preponderant (Figueroa-Espinoza & Rouau, 1998; Schooneveld-Bergmans et al., 1999; Vansteenkiste et al., 2004; Carvajal-Millan et al., 2005a-c; Saulnier et al., 2007). The involvement of a trimer of ferulic acid (4-O-8', 5'-5''-dehydrotriferulic acid) in laccase cross-linked wheat or maize bran AX has been reported (Carvajal-Millan et al., 2005a, 2007). In addition to covalent cross-links (di-FA, tri-FA), the involvement of physical interactions between AX chains was suggested to contribute to the AX gelation and gel properties (Vansteenkiste et al., 2004; Carvajal-Millan et al., 2005b). AX gels have an interesting technological potential as they are mostly stabilized by covalent linkages, which make them stable upon heating and exhibit no syneresis after long time storage (Izydorczyk & Biliaderis, 1995; Vansteenkiste et al. 2004). The purpose of this research was to study the swelling and microstructure of Tacupeto F2001 AX gels.

2   Materials and Methods

2.1   Materials

Arabinoxylans (AX) were extracted from the endosperm of a spring wheat variety (Tacupeto F2001). Tacupeto F2001 bread wheat variety was kindly provided by a wheat milling industry in Northern Mexico (Molino La Fama). AX from wheat endosperm were obtained as reported by Marquez-Escalante et al. (2013). AX presented an A/X ratio of 0.67, a viscosimetric molecular weight

748 kDa, an intrinsic viscosity ($[\eta]$) value of 3.26 dL/g and a FA content of 0.53 mg FA/g AX. Laccase (E.C. 1.10.3.2) from *Trametes Versicolor* and other chemical products were purchased from Sigma Chemical Co. (St Louis, Mo, USA).

2.2   Gelation

AX solution at 2% (w/v) was prepared in 0.1 M sodium acetate buffer pH 5.5. The formation of the AX gel was followed using a strain-controlled rheometer (Discovery HR-3 rheometer, TA instruments) in oscillatory mode as follows as reported before [13]. Cold (4 °C) solutions of 2% (w/v) AX were mixed with laccase (1.675 nkat per mg AX) and immediately placed in the cone and plate geometry (5.0 cm in diameter, 0.04 rad in cone angle) maintained at 4 °C. AX gelation kinetic was monitored at 25 °C for 2 h by following the storage (G') and loss (G'') modulus. All measurements were carried out at a frequency of 1 Hz and 5% strain (linearity range of viscoelastic behavior). Frequency sweep and strain sweep were carried out at the end of the network formation at 25 °C. Reverse phase high-performance liquid chromatography (RP-HPLC) was used to quantify FA, di-FA and tri-FA contents in AX gels after a deesterification step, as described by Rouau *et al.* (2003) and Vansteenkiste *et al.* (2004).

2.3   Swelling and structure

AX gels were allowed to swell in 20 mL of 0.02 % (w/v) sodium azide solution to prevent microbial contamination. During 16 h the samples were taken out, blotted and weighed. The equilibrium swelling was reached when the weight of the samples changed by no more than 3 % (+ 0.03 g). After weighing, a new aliquot of sodium azide solution was added to the gels. Gels were maintained at 25 °C during the test. The swelling ratio (q) was calculated as

$$q = (Ws - Wd)/Wd \qquad (1)$$

where Ws is the weight of swollen gels and Wd is the weight of WBAX in the gel (Carvajal-Millan *et al.*, 2005b).

From swelling measurements, the molecular weight between two cross-links (Mc) was calculated using the classic Flory & Rehner (1943) modified by Peppas & Merrill (1976) analysis for gels where the cross-links are introduced in solution. After Mc calculation the average mesh size (ξ) and the cross-linking density (ρc) in the AX gels were calculated as reported by Carvajal-Millan *et al.* (2005b).

2.4 Microstructure

AX gels were frozen by liquid nitrogen immersion and then lyophilized at −37 °C/0.133 mbar overnight in a Freezone 6 freeze drier (Labconco, Kansas, MO, USA). For scanning electron microscopy lyophilized AX gels were disposed on aluminum stand employed conductive self-adhesive carbon label. Samples were examined without coating at low voltage (1.8 kV) in an EVO LS10 Zeiss scanning electron microscope (Carl Zeiss, Oberkochen. Germany). SEM images were obtained in secondary electrons modes.

2.5 Statistical analysis

Chemical determinations were made in triplicates and the coefficients of variation were lower than 7%. Small deformation measurements were made in triplicates and the coefficients of variation were lower than 9%. Rehydration tests were made in triplicates, coefficients of variation were lower than 10%. All results are expressed as mean values.

3 Results and discussion

3.1. Gelation

The gelation of 2% (w/v) AX solution over time was rheologically investigated by a small amplitude oscillatory shear. The storage (G′) and loss (G″) modulus increased against time but with G′ increasing faster than G″. At ~9 minutes, there was a cross-over where G′ becomes greater than G″ which is associated with the gelation time. After 2 h of laccase exposure the G′ and G″ values were 54.6 and 7.3 Pa, respectively (Figure 1), being close to those reported before for other wheat AX gels at the same polysaccharide concentration (Carvajal-Millan *et al.*, 2005c). The tan δ (G″/G′) value at the end of gelation was 0.12, indicating the formation of a gel as according to Ross-Murphy (1984) the gel point can be detected when G′ becomes greater than G″ (i.e., when tan δ becomes just less

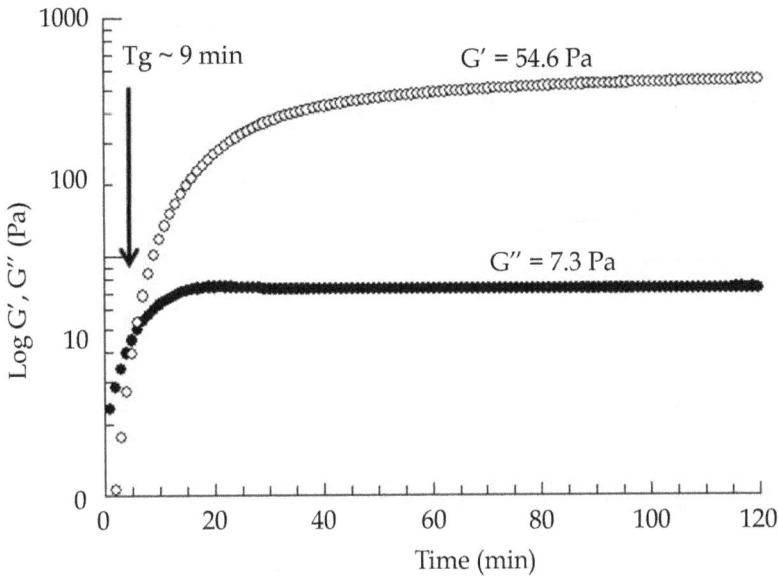

Figure 1. Rheological (G'●, G"O) kinetics of 2% (w/v) AX solution gelation.

than 1). The di-FA content of the AX gels was 0.12 µg/mg AX; the tri-FA was present only in low amount (0.003 µg/mg AX). The mechanical spectra and the strain sweep of AX gels after 2 h of laccase exposure (Figures 2 and 3, respectively) were typical of solid-like materials with a linear G' independent of frequency and strain and G" much smaller than G' and dependent of frequency (Ross-Murphy, 1984).

3.2   Swelling

The swelling of AX gels was followed as a function of time. During 10 hours AX gels registered a constant increase in water uptake (Figure 4). The swelling ratio (q, g water/g AX) at equilibrium swelling of AX gels in 10 h was 20, which is in the range reported for AX gels (Carvajal-Millan *et al.*, 2005b-c; Berlanga-Reyes *et al.* (2009a).

The FA content in the AX sample used in the present study was constant. In a previous study (Carvajal-Millan *et al.*, 2005b) reported that the q value of gels at 1% (w/v in AX) increased from 134 to 223 g water/g AX as the initial FA

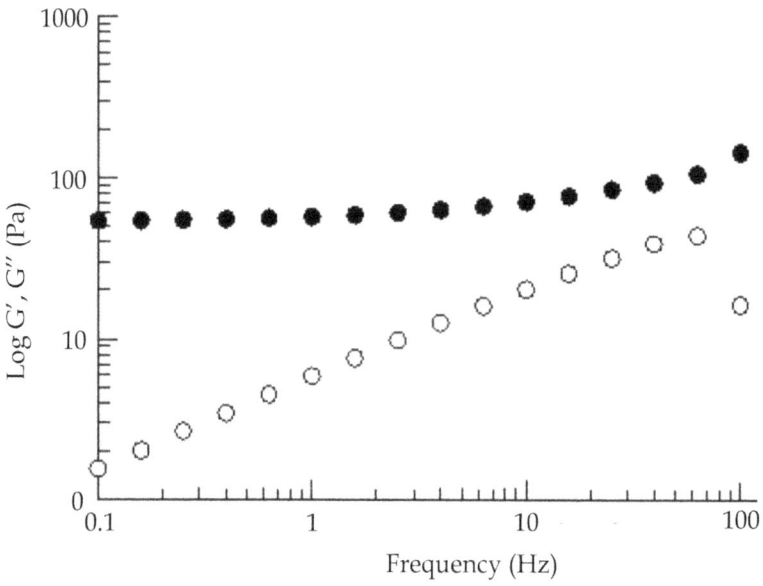

Figure 2. Mechanical spectrum of AX gel at 2% (w/v). (G'●, G''O).

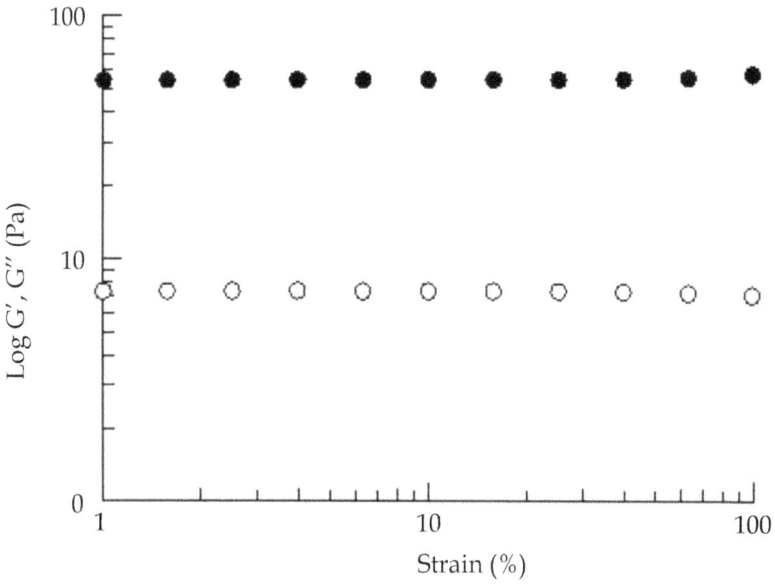

Figure 3. Strain sweep of AX gel at 2% (w/v). (G'●, G''O).

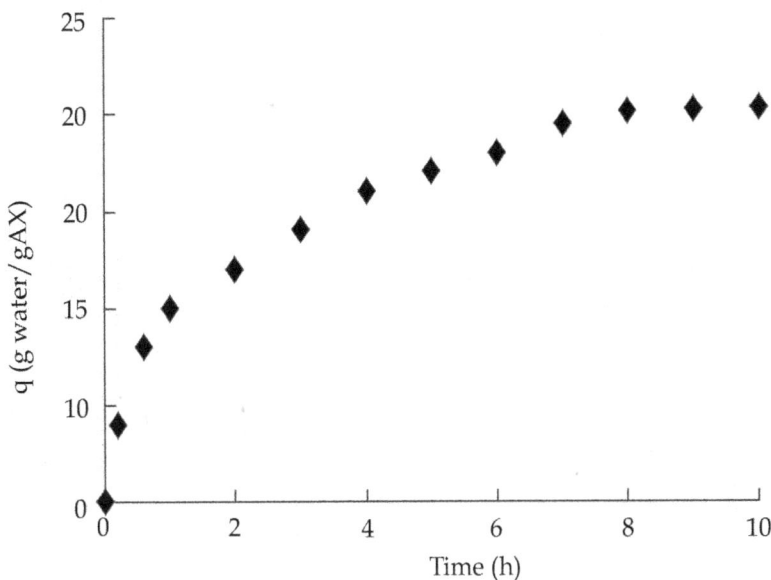

Figure 4. Swelling of AX gel in sodium azide at 0.02% (w/v).

content of AX decreased from 2.3 to 1.6 µg/mg AX. The increase in water uptake accompanying the decrease in AX initial FA content can be explained in terms of a decrease in covalent cross-links (di-FA and tri-FA) in the AX gel. From swelling measurements, Mc, ρc and ξ values of AX gels were calculated and found to be 36 x103 g/mol, 40 x10–6 mol/cm3 and 221 nm, respectively. Similar Mc, ρc and ξ values have been reported in laccase induced AX gels (Carvajal-Millan *et al.*, 2005b; Berlanga-Reyes *et al.*, 2009a).

3.3 Microscopy

The microstructure of the lyophilized AX gel was investigated by scanning electron microscopy (SEM) (Figure 5A and 5B). The AX gel presents many connections and resembles an imperfect honey comb. This material presents average inner dimensions of the cell of approximately 30 × 50 µm (Figure 5 B). Higher cell dimensions were recently reported by Morales-Ortega *et al.* (2013) in lyophilized AX gels (200 x 400 µm), this difference could be related to the method used to freeze AX gels before lyophilization. In the present study AX gels were frozen by immersion in liquid nitrogen (fast congelation) while

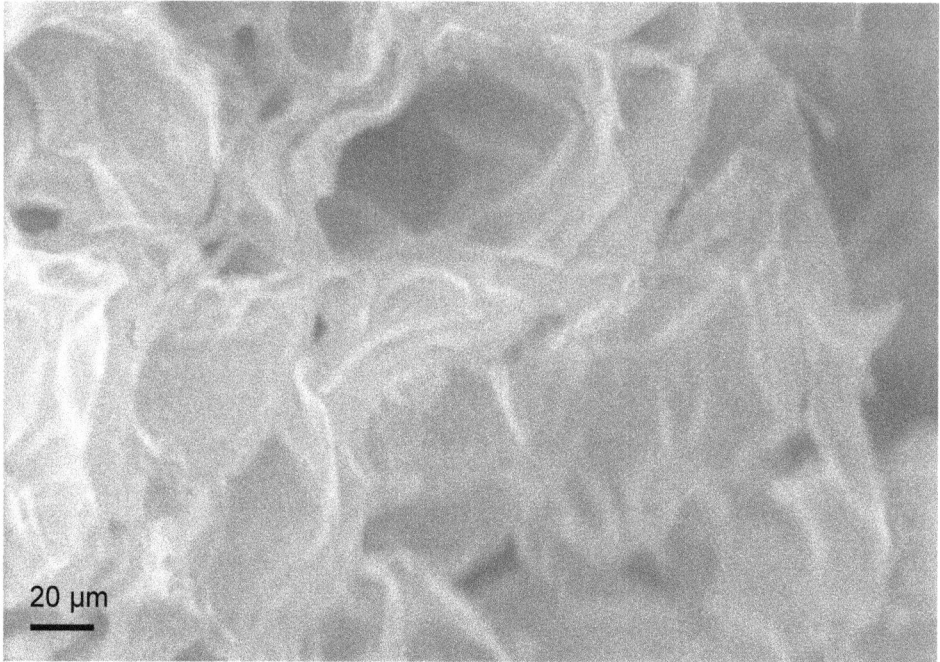

(A)

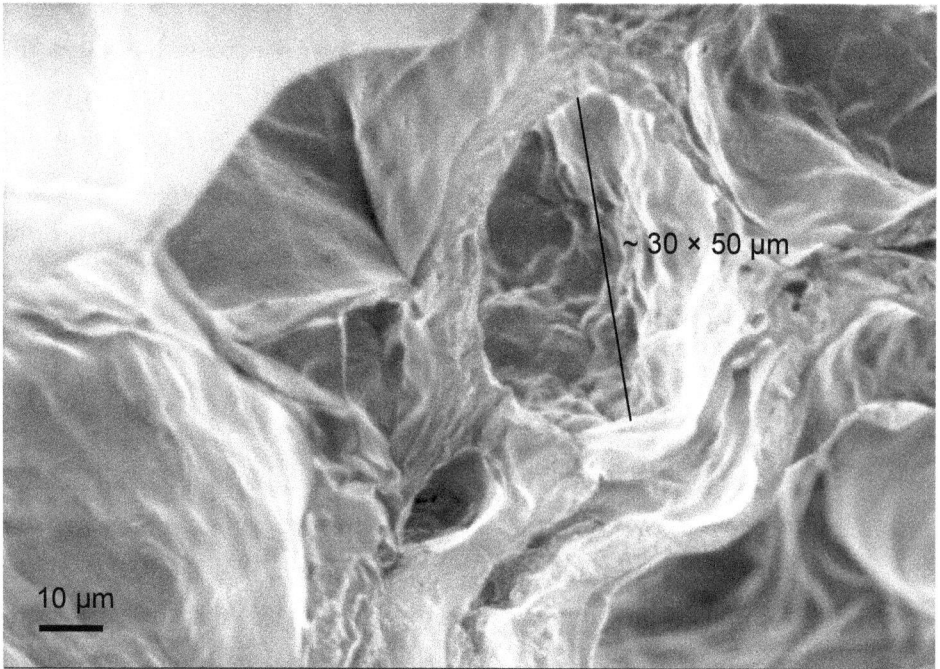

(B)

Figure 5. SEM images of AX gel at (A), 1000× magnification and (B) 2500× magnification.

Morales-Ortega *et al.* (2013) reported AX gel congelation at -20 °C for several hours (slow congelation). It is generally assumed that a rapid temperature drop results in a better preserved structure (i.e. finner ice crystals, smaller changes in drip loss amount). Several studies indicate that an important aspect of achieving a high quality frozen material particularly with high water content such as gels is the knowledge of rate of temperature change (freezing rate) and its relationship to nature and size of ice crystals formation (Da-Wen Sun, 2006; Heldman and Taylor, 1998; Zhu *et al.*, 2005). The microstructural characteristics of AX gels prepared in the present study were similar to those reported by Paës and Chabbert (2012) with AX gels appearing as a mix of sheets and fine strings in which diffusion pathways of free molecules are not straightforward. These microstructural characteristics could be of interest for the development of AX gels as delivery systems of small size bioactive compounds or microorganisms.

4 Conclusions

Tacupeto F2001 AX gels present rheological, swelling and microstructural characteristics similar to those previously reported for other wheat varieties. By using immersion in liquid nitrogen (fast congelation) before lyophilization Tacupeto F2001 AX gels presents cells average inner dimensions lower than those reported by using slow congelation. Rheological, swelling and microstructural characteristics of Tacupeto F2001 AX gels could be of interest for the development of AX gels as delivery systems of bioactive compounds (i.e lycopene, therapeutics proteins, enzymes) or microorganisms (i.e probiotics), allowing alternative uses for this wheat cultivar.

Acknowledgements

This research was supported by Fondo Sectorial de Investigación en Salud y Seguridad Social SSA/IMSS/ISSSTE-CONACYT, Mexico (Grant 179746 to E. Carvajal-Millan). The authors are pleased to acknowledge Valérie Micard, Cécile Barron and Aurélie Putois (SupAgro, INRA, Montpellier, France) for the di-FA and tri-FA acids analysis and to the Laboratory of microscopy (CIAD) for SEM images.

References

Berlanga-Reyes, C.M., Carvajal-Millan, E., Caire Juvera, G., Rascón-Chu, A., Marquez-Escalante, J.A., Martínez-López, A.L. (2009a). Laccase induced maize bran arabinoxylan gels: structural and rheological properties. Food Science and Biotehnology, 18, 1027–1029.

Berlanga-Reyes, C.M., Carvajal-Millan, E., Lizardi-Mendoza, J., Rascón-Chu, A., Marquez-Escalante, J.A., Martínez-López, A.L. (2009b). Maize arabinoxylan gels as protein delivery matrices. Molecules, 14, 1475–1482.

Carvajal-Millan, E., Guigliarelli, B., Belle, V., Rouau, X., Micard, V. (2005a). Storage stability of laccase induced arabinoxylan gels. Carbohydrate Polymers, 59, 181-188.

Carvajal-Millan, E., Landillon, V., Morel, M.H., Rouau, X., Doublier, J.L., Micard, V. (2005b). Arabinoxylan gels: impact of the feruloylation degree on their structure and properties. Biomacromolecules, 6, 309-317.

Carvajal-Millan, E., Guilbert, S., Morel, M.H., Micard V. (2005c). Impact of the structure of arabinoxylan gels on their rheological and protein transport properties. Carbohydrate Polymers, 60, 431-438.

Carvajal-Millan, E., Rascón-Chu, A., Márquez-Escalante, J., Ponce de León, N., Micard, V., Gardea, A. (2007). Maize bran gum: extraction, characterization and functional properties. Carbohydrate Polymers, 69, 280-285.

Da-Wen Sun, 2006. Handbook of Frozen Food Processing and Packing. CRC Press Taylor & Francis Group, Boca Raton, 723 p.

Figueroa-Espinoza, M.C., Rouau, X. (1998). Oxidative cross-linking of pentosans by a fungal laccase and horseradish peroxidase: mechanism of linkage between feruloylated arabinoxylans. Cereal Chemistry, 75, 259-265.

Flory, P.J., Rehner, J. (1943). Statistical mechanics of cross-linked polymer networks. II. Swelling. Journal of Chemical Physics, 11, 521-526.

Geissman, T., Neukom, H. (1973). On the composition of the water-soluble wheat flour pentosans and their oxidative gelation. Lebensmittel-Wissenschaft und Technologie, 6, 59-62.

Heldman, D.R., Taylor, T.A., 1998. Modelling of food freezing. In: Erickson, M.C., Hung, Y.-C. (Eds.), Quality of Frozen Foods. Chapman Hall, London, pp. 51e64.

Hoseney, R.C., Faubion, J.M. (1981). A mechanism for the oxidative gelation of wheat flour water-soluble pentosans. Cereal Chemistry, 58, 421-424.

Izydorczyk, M.S.; Biliaderis, C.B. (1995). Cereal arabinoxylans: Advances in structure and physicochemical properties. Carbohydrate Polymers, 28, 33–48.

Martinez-Lopez, A.L., Carvajal-Millan, E., Miki-Yoshida, M., Alvarez-Contreras, L., Rascon-Chu, A., Lizardi-Mendoza, J., Lopez-Franco., Y. (2013) Arabinoxylan microspheres: structural and textural characteristics. Molecules, 18, 4640-4650.

Marquez-Escalante, J., Carvajal-Millan, E., Miki-Yoshida, M., Alvarez-Contreras, L., Toledo-Guillén, A.R., Lizardi-Mendoza, J., Rascón-Chu, A. (2013). Water extractable arabinoxylan aerogels prepared by supercritical CO_2 drying. Molecules, 18, 5531–5542.

Morales-Ortega, A., Carvajal-Millan, E., López-Franco, Y., Rascón-Chu, A., Lizardi-Mendoza, J., Torres-Chavez, P., Campa-Mada, A. (2013). Characterization of Water Extractable Arabinoxylans from a Spring Wheat Flour: Rheological Properties and Microstructure. Molecules, 18, 8417-8428.

Morales-Ortega, A., Carvajal-Millan, E., Brown-Bojorquez, F., Rascón-Chu, A., Torres-Chavez, P., López-Franco, Y., Lizardi-Mendoza, J., Martinez-Lopez, A.L., Campa-Mada, A. (2014). Entrapment of prebiotics in water extractable arabinoxyllan gels: rheological and microstructural characterization. Molecules, 19, 3628-3637.

Niño-Medina, G., Carvajal-Millan, E., Rascon-Chu, A., Márquez-Escalante, J.A., Guerrero V., Salas-Muñoz, E. (2010). Feruloylated arabinoxylans and arabinoxylan gels: Structure, sources and applications. Phytochemestry Review, 9, 111–120.

Paes, G., Chabbert, B. (2012). Characterization of Arabinoxylan/cellulose nanocrystals gels to investigate fluorescent probes mobility in bioinspired models of plant secondary cell wall. Biomacromolecules, 13, 206–214.

Peppas, N.A., Merrill, E.W. (1976). Poly(vinyl alcohol) hydrogels: reinforcement of radiation-crosslinked networks by crystallization. Journal of Polymer Science, 14, 441-457.

Ross-Murphy, S.B. (1984). Rheological methods. In H. W. S. Chan, Biophysical Methods in Food Research (pp.138-199). Oxford: Blackwell.

Saulnier, L., Sado, P.E., Branlard, G., Charmet, G.; Guillon, F. (2007). Wheat arabinoxylans: Exploting variation in amount and composition to develpop enhanced varieties. Journa of Cereal Science, 46, 261–281.

Schooneveld-Bergmans, M.E.F., Dignum, M.J.W., Grabber, J.H., Beldman, G., Voragen, A.G.J. (1999). Studies on the oxidative cross-linking of feruloylated arabinoxylans from wheat flour and wheat bran. Carbohydrate Polymers, 38, 309-317.

Smith, M.M., Hartley, R.D. (1983). Ocurrence and nature of ferulic acid sudstitution of cell-wall polysaccharides in graminaceous plants. Carbohydrate Research, 188, 65–80.

Vansteenkiste, E., Babot, C., Rouau, X., Micard, V. (2004). Oxidative gelation of feruloylated arabinoxylan as affected by protein. Influence on protein enzymatic hydrolysis. Food Hydrocolloids, 18, 557-564.

Zhu, S., Ramaswamy, H.S., Le Bail, A. (2005). Ice-crystal formation in gelatine gel during pressure shift versus conventional freezing. Journal Food Engineering, 66, 69-76.

Lapachol and Its Derivatives as Potential Drugs for Cancer Treatment

Júlio César Nepomuceno[1,2]

1 Introduction

Natural products extracted from plants have significantly contributed to the development of various medications used clinically in traditional medicine (Newman *et al.*, 2003). Prospection of biological materials in plants, in areas of wide biodiversity in tropical and subtropical regions, provides chemical diversity for obtaining products for the development of new drugs (Kinghorn *et al.*, 2003). Natural products have been widely used in treatment of diseases in popular medicine. Based on this tradition the use of medication produced from plants for the treatment of cancer has become common. Between 1983 and 1994, approximately 62% of commercial drugs used to treat cancer were derived from natural sources (Ravelo *et al.*, 2004).

Tabebuia impetiginosa, also known as "ipe" (Pau d'arco), is one of the various species of plants tested for chemical substances. The active chemical compound found in this plant is naphthoquinone, also known as lapachol. Lapachol (4-hydroxy-3-(3-methylbut-2-enyl)naphthalene-1,2-dione) is a natural quinone that can be isolated from many species of Bignoniaceae, family, specifically those of the genus *Tabebuia* (*Tabebuia aurea, Tabebuia impetiginosa, Tabebuia ochracea*), from the Brazilian Cerrado. It is found in Brazil, and is commonly known as "ipê amarelo" (Oliveira, 2000). A number of studies have examined this natural quinine, including its anticancer, antiviral, antimicrobial, anti-

[1] Instituto de Genética e Bioquímica, Universidade Federal de Uberlândia, Brasil
[2] Laboratório de Citogenética e Mutagênese, Centro Universitário de Patos de Minas, Brasil

inflammatory, and antimalarial effects, as well as its significant effect on *Trypanosoma cruzi* (Carvalho *et al.*, 1998; Grazziotin *et al.*, 1992). This quinone showed significant in vivo anti-tumor activity in several early mouse models (Rao *et al.*,1968), since then progressing to clinical trials by the National Cancer Institute (NCI) in the 1970's. However, in 1974, the NCI concluded that the high concentrations required for efficient chemotherapy in human cancer treatment, unfortunately also gave rise to extremely toxic side-effects, thereby justifying its rejection (Suffness & Douros, 1980; Castellanos *et al.*, 2009).

Quinones are classified by the aromatic moieties present in their structure and naphthoquinone constitutes the naphtalenic ring (Silva *et al.*, 2003). The naphthoquinones are a class of compounds having cytotoxic properties that can be advantageous for treating cancer. Two essential mechanisms are linked to the effects of naphthoquinone, oxidative stress and nucleophilic alkylation (Bolton *et al.*, 2000). These substances are able to accept electrons and generate reactive oxygen species (hydrogen peroxide ($H2O2$), superoxide anion radical ($O2$•) and hydroxyl radical (HO•), whose oxidative effects could explain the cytotoxicity produced by these compounds (Boveris *et al.*, 1978; Silva *et al.*, 2003). Bolton *et al.*, (2000) suggested that quinones are highly reactive molecules and can reduce the redox cycle by using semi-quinine radicals, generating reactive oxygen species (ROS) that include superoxide radicals, peroxide radicals, hydrogen peroxide, and hydroxyl radicals. ROS production can cause severe oxidative cell stress, forming oxidative macromolecule cells, affecting lipids, proteins and DNA.

Lapachol proved to be a vitamin K-antagonist antigen, thus possibly targeting vitamin K-dependent reactions (Dinnen & Ebisuzaki, 1997), besides also being bio-activated by P450 reductase to reactive species which promote DNA scission, through redox cycling with generation of free radicals (Kumagai *et al.*, 1997). The enzyme responsible for bioactivating lapachol, thereby leading to the generation of ROS capable of causing DNA damage, was unknown. An immunoinhibition study with antibodies against cytochrome P450 reductase (P450R) revealed that P450R was a pre- dominant enzyme in catalyzing the one-electron reduction of lapachol (Kumagai *et al.*, 1997). Generation of reactive oxygen species, superoxide anion radical and hydroxyl radical during the metabolism of lapachol by P450 reductase, was confirmed by acetylated cytochrome reduction assay in the absence and presence of Cu, Zn-SOD (Superoxide Dismutase), and electron spin resonance (ESR) studies (Kumagai *et al.*, 1997).

Recently, a new series of compounds via the molecular hybridization of naphthoquinones (such as naturally occurring lapachol) with pterocarpans (naturally occurring isoflavonoids) was designed and synthesized. These hybrid pterocarpanquinones display in vitro antineoplastic and antiparasitic activities. Therefore, the purpose of this chapter is to review studies in humans and animals in order to evaluate the use of lapachol and its derivatives as a therapeutic intervention in cancer patients.

2 Pau d'arco

Approximately two-thirds of the biological diversity of the world is found in tropical zones, mainly in developing countries. Brazil is considered the country with the greatest biodiversity on the planet, forming several biomes (Vieira, 1999).

The Cerrado is the second largest ecological dominion of Brazil, where a continuous herbaceous stratum is joined to an arboreal stratum, with variable density of woody species. The Cerrado covers a surface area of approximately 25% of Brazilian territory, and around 220 species found in the Cerrado are reportedly used in traditional medicine (Vieira & Martins, 1998). Plants have played a significant role in maintaining human health and improving the quality of human life for thousands of years, and have served humans well as valuable components of seasonings, beverages, cosmetics, dyes, and medicines. The World Health Organization estimated that 80% of the earth's inhabitants rely on traditional medicine for their primary health care needs, and most of this therapy involves the use of plant extracts or their active components (Craig, 1999).

Pau d'arco is a huge canopy tree native to the Amazon rainforest and other tropical parts of South and Latin America. It grows to 30 m high and the base of the tree can be 2 – 3 m in diameter. The *Tabebuia* genus includes about 100 species of large, flowering trees that are common in South American cities's landscapes due to their beauty. The tree also is popular with timber loggers — its this quinone showed significant in vivo anti-tumor activity in several early mouse models (Rao *et al.*,1968). Pau d'arco wood is widely used in the construction of everything from houses and boats to farm tools. The common name pau d'arco (as well as its other main names of commerce, ipê roxo and lapacho) is used for several different species of Tabebuia trees that are used interchangeably in herbal medicine systems. *T. impetiginosa* is known for its attractive purple flowers and often is called "purple lapacho" (http://www.rain-tree.com/paudarco.htm#.UwOiRPldUmu).

Tabebuia impetiginosa, also known as "ipe", is one of the various species of plants tested for chemical substances. The active chemical compound found in this plant is naphthoquinone, also known as lapachol. Quinones are classified by the aromatic moieties present in their structure and naphthoquinone constitutes the naphtalenic ring (Silva *et al.*, 2003).

Pau d'arco (*Tabebuia avellanedae*) (Figure 1), where it has been used to treat a wide range of conditions, including pain, arthritis, inflammation of the prostate gland (prostatitis), fever, dysentery, boils and ulcers, and various cancers. As early as 1873, there are reports of medicinal uses of Pau d'arco. Scientists have identified two active chemicals in pau d'arco. These chemicals are called naphthoquinones: lapachol and beta-lapachone. In lab tests, these chemicals kill some bacteria, fungi, viruses, and parasites. They also have anti-inflammatory properties. But no one knows whether they will have the same effects when humans take them, and the usual dose required would have severe, toxic side effects. Pau d'arco is sometimes used for the following conditions, although there is no evidence it works: Candidiasis (a vaginal or oral yeast infection); Herpes simplex; virus influenza; Parasitic diseases, such as schistosomiasis; Bacterial infections, such as brucellosis and Cancer (http://umm.edu/health/medical/altmed/herb/pau-darco).

Tea made from pau d'arco is thought to have been used by the ancient Incas and natives of the South American rain forests, who took it to cure disease and as a tonic to strengthen the body and improve overall health. Caribbean folk healers reportedly use the leaf and the bark to treat backaches, toothaches, and sexually transmitted diseases. The native tribes of Brazil used the tree to make bows for hunting. When the Portuguese colonized Brazil, they named the tree pau d'arco, which means "bow stick." The herb remains a popular Brazilian folk remedy. Pau d'arco is available as a capsule, tablet, salve, liquid extract, powder, and tea from health food stores many drugstores, and Internet sources. Recommended dosage varies by manufacturer. When making tea, practitioners

Figure 1: *Tabebuia avellanedae.* (Hussain *et al.*, 2007).

say the bark must be boiled or simmered for at least 8-10 minutes to release the active ingredients, which do not dissolve easily in water (American Cancer Society, 2014).

According to Taylor (2005), unfortunately, its popularity and use have been controversial due to varying results obtained with its use. For the most part, these seem to have been caused by a lack of quality control—and confusion as to which part of the plant to use and how to prepare it. Many species of *Tabebuia*, as well as other completely unrelated tree species exported today from South America as "pau d'arco," have few to none of the active constituents of the true medicinal species. Pau d'arco lumber is in high demand in South America. The inner bark shavings commonly sold in the U.S. are actually by-products of the timber and lumber industries. Even mahogany shavings from the same sawmill floors in Brazil are swept up and sold around the world as "pau d'arco" (due to the similarity in color and odor of the two woods). In 1987, a chemical analysis of 12 commercially-available pau d'arco products revealed only one product containing lapachol—and only in trace amounts. As lapachol concentration typically is 2 – 7% in true pau d'arco, the study surmised that the products were not truly pau d'arco, or that processing and transportation had damaged them. Most pau d'arco research has centered on the heartwood of the tree.

3   Lapachol

Lapachol is a naphthoquinone that was first isolated by E. Paterno from *Tabebuia avellanedae* (Bignoniaceae) in 1882. Originally isolated from species of the Bignoniaceae family, lapachol can also be found in other families such as Verbenaceae, Proteaceae, Leguminosae, Sapotaceae, Scrophulariaceae, and Malvaceae (Hussain *et al.*, 2007) (Table 1). Lapachol (LAP), 4-hydroxy-3-(3-methylbut-2-enyl)naphthalene-1,2-dione (Figure 2), is a naphthoquinone found in many vegetable species related to the Bignoniaceae family, specifically those belonging to the plant genus *Tabebuia* (*Tabebuia aurea, Tabebuia impetiginosa, Tabebuia ochracea*) found in the Cerrado (Brazil), commonly known as "pau d'arco".

In 1956, Lima *et al.*, (1956), found at the core of pau d'arco purple bow, *Tabebuia sp.*, the presence of a compound of yellow color with antimicrobial activity, especially for the genus *Brucella*, and that in the course of studies chemical reached the conclusion that it was lapachol.

4   Beta Lapachone

β-Lapachone (3,4-dihydro-2,2-dimethyl-2H-naptho[1,2-b]pyran-5,6-dione) (see Figure 3) is a naturally occurring quinone derived from the lapacho tree, (*Tabeuia avellanedae*) native to Central and South America. The synthesis and chemistry of β-lapachone and related compounds was initially investigated in the late 19th and early 20th centuries by the chemist Samuel Hooker. Although some clinical studies on the bioactivity of naphthoparaquinones were reported early in 1940, it was Docampo's report of β-lapachone's potential as an anti-trypanosomal agent in 1977 that revitalized the study and characterization of this unique natural compound (Hussain, 2007).

Family	Species
Bignoniaceae	*Tabebuia flavescens* Benth - & Hook. F. ex. Griseb.
	T. guayacan Hemsl.
	T. avellanedae Lor. ex Griseb.
	T. serratifolia (Vahl.) Nichols
	T. rosa
	T. bata (E. Mey) Sandw
	T. pentaphylla (Linn) Hemls.
	T. heptaphylla
	Haplophragma adenophylum
	Heterophragma adenophylum
	Kigelia pinnata
	Phyllarthron comorense
	Radermachera sinica
	Paratecoma peroba (Record) Kuhlm
	Tecoma araliaceae DC
	T. undulata
	Stereospermum suaveolens DC.
	S. kunthianum
	Zeyhera digitalis
	Z. tuberculosa
	Millingtonia hortensis Linn
	Stereospermum tetragonum DC.
	Stereospermumpersonatum
	Catalpa longíssima
	Cybistax antisyphilitica
	Macfadyena ungis-cati
	Melloa quadrivalvis
	Newbouldia laevis
Verbenaceae	*Tectona grandis* L. fil
	Avicennia tomentosa Jacq
	Avicenia officinalis
Proteaceae	*Conospremum teretifolium* R. Br.

Continued on next page...

…Continued from previous page

Leguminosae	*Diphysa robinoide* Bent
Sapotaceae	*Bassia latifolia*
Malvaceae	*Hibiscus tiliaceus*
Scrophulariaceae	*Paulownia kawakamii*

Table 1: Occurrence of lapachol in families and species. Data from: Hussain *et al., (2007)*.

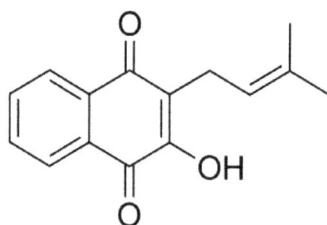

Figure 2: Chemical structure of lapachol.

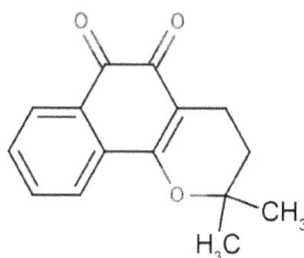

Figure 3: Chemical structure of Beta lapachone.

Studies demonstrated that β-lapachone can directly target DNA topoisomerases and inhibit their activity, which results in cytotoxicity (Lee *et al.*, 2005). The DNA supercoiling is a precisely regulated process that influences DNA replication, transcription, and packaging. The DNA topoisomerases are enzymes that modulate the topological state of DNA (Esteves-Souza *et al.*, 2007)

Beta-lapachone is bioactivated by NAD(P)H:quinone oxidoreductase-1 (NQO1), creating a futile oxidoreduction that generates high levels of superoxide. In turn, the highly reactive oxygen species (ROS) interact with DNA, thereby causing single-strand DNA breaks and calcium release from endoplasmic reticulum (ER) stores. Eventually, the extensive DNA damage causes hyperactivation of poly(ADP-ribose) polymerase-1 (PARP-1), an enzyme facilitating DNA repair, accompanied by rapid depletion of NAD+/ATP nucleotide levels. As a result, a caspase-independent and ER-stress induced mu-calpain- mediated cell death occurs in NQO1-overexpressing tumor cells. NQO1, a flavoprotein and two-electron oxidoreductase, is overexpressed in a variety of tumors (National Cancer Institute, 2014).

This ortho-naphthoquinone has been widely studied, mainly by presenting selective effects on tumor cell lines compared to normal strains. The indication that β-lapachone can lead to formation of oxygen free radicals may be your primary strategy for toxicity against human tumors, however, this effect usually leads to formation of intermediates that are highly toxic to biological systems (Asche, 2005).

5   Pterocarpanquinones

Over the last years the Laboratory of Bioorganic Chemistry of the Federal University do Rio de Janeiro (LQB-UFRJ®) has contributed to the identification of new bioactive compounds, with the inspiring source molecular scaffolds present in natural sources, like cumestanos and Pterocarpans. This evolution began with the identification of antiophidic cumestanos, passing Pterocarpans with wide spectrum of biological activity until the discovery of pterocarpanoquinones LQB-118. (Costa, 2009).

The pterocarpanquinones comprise a new group of antineoplasic and antiparasitic prototypes. The first compounds of this series were designed and synthesized by da Silva et al., (2003) some years ago. According to Costa et al., (2012) these compounds showed antineoplasic activity in various tumor cell lines. Pterocarpanquinones is a result from molecular hybridization of pterocarpans and the naphtoquinone with anti-tumoral activity lapachol [2-hydroxy-3-(3-methyl-2-butenyl)-1,4-naphthoquinone] (Maia et al., 2011).

Fourteen compounds, bearing a new molecular architecture (pterocarpanquinone) designed by molecular hybridization of 4-hydroxymaackian, lapachol and calafungin, were synthesized and evaluated as anticancer and antiparasitic. These pterocarpanquinones showed to be more active and bioselective than LQB-79 (pterocarpan) on leukemia, breast and lung cancer cell lines, as well as on Plasmodium falciparum and Leishmania amazonensis in culture. One among these compounds, LQB-118 was active in mice infected (footpad) by *Leishmania amazonensis*.

Pterocarpanquinone LQB-118 (Figure 4) is the most promising compound of this series. The antineoplasicaction of LQB-118 on chronic myeloid leukemia was further studied and this compound led to significant apoptosis rate in cells from patients with chronic myeloid leukemia in treatment in the National Institute of Cancer (INCA) in Rio de Janeiro (Costa *et al.*, 2012).

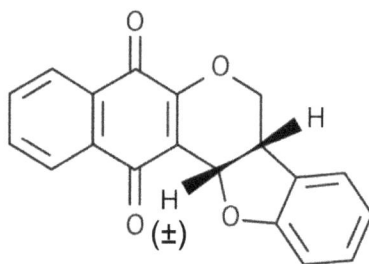

Figure 4: Chemical structure of ± LQB 118 (pterocarpanquinone).

LQB-118 was synthesized through a new palladiumcatalyzed oxyarylation (oxa-Heck) reaction of chromenquinone 1 by ortho-iodophenol 2 in the absence of ligands (Figure 5). After flash chromatography, the product was obtained as a yellow solid in 39% yield (Costa *et al.*, 2012).

6 Studies of Lapachol and its Derivatives in Cancer Treatment

Despite all the technological advancement in the treatment of cancer disease it is still a challenge and the cases of healing are still rare. Natural products and

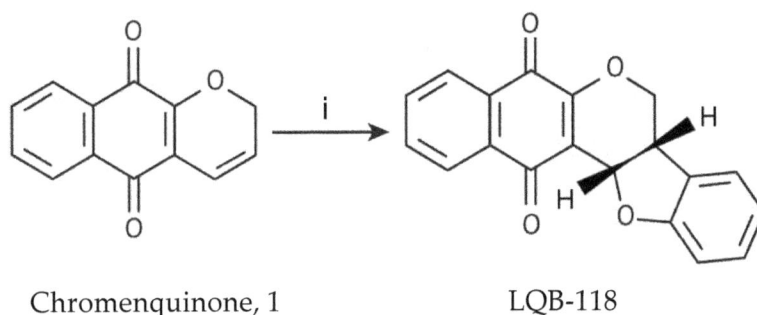

Chromenquinone, 1   LQB-118

Figure 5: Synthesis of LQB-118. i: orho-iodophenol, 2, Pd(OAc)$_2$, Ag$_2$CO$_3$, acetone, reflux, 39%.

chemical modification of antitumor substances are among the most important strategies used in the search for new anticancer drugs (Esteves-Souza *et al*, 2008).

Naturally occurring quinones and their analogs are important sources of cytotoxic compounds. Dactinomycin, anthracycline antibiotics (daunorubicin, doxorubicin, idarubicin and mitoxantrone), bleomycins and mitomycin-C have been clinically used for cancer chemotherapy (Gewirtz, 1999; Galm *et al.*, 2005; Wolkenberg & Boger, 2005; Zhang *et al.*, 2006).

6.1 Lapachol

In a 1968 study, lapachol demonstrated highly significant activity against cancerous tumors in rats. The antitumor activity of lapachol was tested against carcinoma 755, leukemia L-1210, P-1534 leukemia, Sarcoma 180, and Walker 256 carcinosarcoma. Of these, only Walker 256 showed sensitivity to lapachol when administered by daily i.p. injection at levels up to those producing limiting toxicity. The activity of lapachol against the Walker 256 intramuscular tumor system was first noted when the compound was administered by the i.p. route. Its antitumor effect was then studies in detail through dosage of implanted animals by a number of parenteral routes and orally and by a variety route on treatment regimens (Rao *et al.*, 1968). Because of the highly significant activity in the Walker 256 intramuscular tumor system (Table 2), especially when the drug is given twice daily by the oral route, and because of the relatively mild signs of

Compounds	Dose mg/kg	YoshidaSarcoma	Walker 256 Tumor
Lapachol	100	82%	50%
	100	76,7%	40,26%
	150	----	51,00%
	160	71,1%	----
	200	82,7%	----
	250	59,68%	----
	300	65,49%	----
Aqueous extract of the cortex of Pau d'Arco	150	85,00 %	80,00%

Table 2: Action Lapachol and its derivatives on the Yoshida sarcoma and Walker 256 tumor.

general toxicity, lapachol was approved by the Cancer Chemotherapy National Service Center (CCNSC) for human clinical trials (Rao *et al.*, 1968).

However, in 1974, the National Cancer Institute (NCI) concluded that the high concentrations required for efficient chemotherapy in human cancer treatment, unfortunately also gave rise to extremely toxic side-effects, thereby justifying its rejection (Suffness & Douros, 1980; Castellanos *et al.*, 2009).

In a small study in 1980 with nine patients with various cancers (liver, kidney, breast, prostate and cervix), pure lapachol demonstrated an ability to shrink tumors and reduce feeling of pain caused by these tumors and achieved complete remissions in three of the patients. It is believed that the antitumor activity of lapachol may be related to its interaction with nucleic acids. Additionally it has been proposed that interaction of the naphthoquinone moiety between base pairs of the DNA helix occurs with subsequent inhibition of DNA replication and RNA synthesis (Hussain *et al.*, 2007).

Studies conducted by Kumagai *et al.*, (1997), Kumagai and Shimojo (2002) demonstrated that oxidative stress induced by lapachol occurs in response to the P450 reductase enzyme, causing changes in the DNA. Lapachol itself does not have a direct effect on DNA, however, when it interacts with the Cytochromes P450 the effect on DNA can be observed.

Balassiano *et al.*, (2005) analyzed the effects of lapachol on a human cancer cell line and evaluated the potential of this substance as an anti-metastatic drug using an *in vivo* assay. The results of this study indicated that lapachol, in the maximal non-toxic concentration for HeLa cells of 400 μg/ml (corresponding to 10^{12} molecules of the drug/cell), induce alterations in the protein profile and inhibit cellular invasiveness, thus representing an important anti-metastatic activity.

Lapachol has also been demonstrated to reduce the number of tumors caused by doxorubicin in *Drosophila melanogaster* heterozygous for the tumor suppressor gene *wts* (Costa *et al.*, 2011). The frequency of reduction of tumors was directly proportional to the concentration of lapachol. At a concentration of 20 mg/mL, there was a reduction of 64% in the frequency of tumors induced by DXR. At a concentration of 40 mg/mL, the reduction was 71% and at a concentration of 60 mg/mL a 76% reduction was noted. There was tumor reduction throughout the body of the animal, concentrated primarily on the wings and body.

6.2   Beta Lapachone

Hussain *et al.*, (2007) in his careful review of lapachol shows that β-Lapachone is an anticancer agent that selectively induces cell death in several human cancer cell lines. However the precise mechanism of β-lapachone cytotoxicity is not yet fully understood. Hussain *et al.*, (2007) reported that β-lapachone treatment delayed cell cycle progression at the G1/S transition, incremented phosphorylation of the Rad53p checkpoint kinase and decreased cell survival in the budding yeast, *Saccharomyces cerevisiae*. Furthermore, β-lapachone induced phosphorylation of histone H2A at serine 129. These checkpoint responses were regulated by Mec1p and Tel1p kinases. Mec1p was required for Rad53p/histone H2A phosphorylation and cell survival following β-lapachone treatment in asynchronous cultures, but not for the G1 delay. The major and vital conclusion of all those findings indicated that β-lapachone activates a Mre11p-Tel1p checkpoint pathway in budding yeast. Given the conservative nature of the Mre11p-Tel1p pathway, these results suggest that activation of the Mre11-Tel1p checkpoint could be of significance for β-lapachone anti-tumor activity.

Hussain *et al.*, (2007) also reported that ablation of tumor colonies has been observed in a wide spectrum of human carcinoma cells in culture after treatment with a combination of β-lapachone and taxol, two low molecular mass compounds. They synergistically induced death of cultured ovarian, breast, prostate, melanoma, lung, colon, and pancreatic cancer cells. This combination therapy has unusually potent antitumor activity against human ovarian and prostate tumor prexenografted in mice.

According to Park *et al., (2005)* there is a synergistic effect between ionizing radiation and β-lapachone. In the study it was observed that the irradiated cancer cells promoted the action of β-lapachone by elevated levels of NQO1 (Naphthoquinone oxidoreductase-1), which is a processing enzyme of this naphthoquinone pro-drug into a substance with effective action.

Cancer of the colon, liver, breast, ovarian, thyroid, adrenal and corneas have higher expressions of NQO1 compared with normal tissues. Although the majority of tumors present an overexpression of this enzyme there is some types of cancer present in the normal levels. For example, cancer of the stomach and the kidney has low concentrations of NQO1 being compared to normal tissues (Belinsky & Jaiswal, 1993; Siegel *et al.,* 2004.). Due to the fact β-lapachone be related in several studies as a pro-drug, the presence of this enzyme becomes very interesting in fighting cancer.

Kung *et al., (2014)* claim that β-lapachone is a promising potential therapeutic drug for various tumors, including lung cancer, the leading cause of cancer-related deaths worldwide. They found that apoptotic cell death induced in lung cancer cells by high concentrations of β-lapachone was mediated by increased activation of the pro-apoptotic factor JNK and decreased activation of the cell survival/proliferation factors PI3K, AKT, and ERK. In addition, β-lapachone toxicity was positively correlated with the expression and activity of NAD(P)H quinone oxidoreductase 1 (NQO1) in the tumor cells. They also found that the FDA-approved non-steroidal anti-inflammatory drug sulindac and its metabolites, sulindac sulfide and sulindac sulfone, increased NQO1 expression and activity in the lung adenocarcinoma cell lines CL1-1 and CL1-5, which have lower NQO1 levels and lower sensitivity to β-lapachone treatment than the A549 cell lines, and that inhibition of NQO1 by either dicoumarol treatment or NQO1 siRNA knockdown inhibited this sulindac-induced increase in β-lapachone cytotoxicity. The authors conclude the sulindac and its metabolites synergistically increase the anticancer effects of β-lapachone primarily by increasing NQO1 activity and expression, and these two drugs may provide a novel combination therapy for lung cancers.

As previously described, NAD(P)H:quinone oxidoreductase (NQO1) activity is the principal determinant of β-Lapachone cytotoxicity. Knowing this, Lambertia *et al.,* (2013) investigated the feasibility of Photodynamic therapy (PDT) to increase the anticancer effect of β-Lapachone by up-regulating NQO1 expression on breast cancer MCF-7c3 cells. Photodynamic therapy (PDT) is a clinically approved and rapidly developing cancer treatment. PDT involves the administration of photosensitizer (PS) followed by local illumination with visible light of specific wavelength. In the presence of oxygen molecules, the light illumination of PS can lead to a series of photochemical reactions and consequently the generation of cytotoxic reactive oxygen species (ROS). It has been reported that β-Lapachone synergistically interacts with ionizing radiation, hyperthermia and cisplatin and that the sensitivity of cells to β-Lapachone is closely related to the activity of NQO1 (Lambertia *et al.,* 2013). According to authors the results lead us to conclude that the synergistic interaction between β-Lapachone and PDT in killing cells was consistent with the up-regulation of NQO1. The combination of β-Lapachone and PDT is a potentially promising modality for the treatment of cancer.

Larsson *et al.,* (2006) studied in vitro drug sensitivity screening using the fluorometric microculture cytotoxicity assay in one human pancreatic carcinoid and two human bronchial carcinoid cell lines. The aim of this study was to investigate drug sensitivity in neuroendocrine tumor cell lines. A total of 18 drugs, including β-lapachone, with different mechanisms of action were tested. These studies indicated that some of the tested compounds viz., β-lapachone, could possibly be used in clinical trials and demonstrate a therapeutic effect in patients suffering with neuroendocrine tumors.

Several studies have reported the importance of antitumor activity of naphthoquinones. The β-lapachone stands out in this field, and their biological activity, justifies continuing studies with naphthoquinones and derivatives such as activators of oxidative stress agents and topoisomerase II inhibitors (Kongkathip *et al.,* 2004).

The naphthoquinones are widely studied, mainly because of the selective effects on tumor cell lines compared to normal strains. Several mechanisms of action of naphthoquinones are related to dose and time dependent. The observed inhibition of cell growth can occur due to the induction of apoptosis, inhibition of topoisomerase II or oxidative stress, among others (Li *et al.*, 2003; Woo & Choi, 2005; Reinicke *et al.*, 2005). Kung *et al.*, (2007) suggest that β-lapachone may have potential as an antiangiogenic drug in human endothelial cells.

Neovascularization is an essential process in tumor development and thus it is conceivable that anti-angiogenic treatment may block tumor growth. In angiogenesis, nitric oxide (NO) is an important factor which mediates vascular endothelial cell growth and migration. Whether β-lapachone can induce endo-thelial cell death or has an anti-angiogenic effect is still uncertain. Kung *et al.*, (2007) investigated the in vitro effect of β-lapachone on endothelial cells, includ-ing the human vascular endothelial cell line, EAhy926, and human umbilical vascular endothelial cells (HUVEC). Kung *et al.*, (2007) demonstrated that NO can attenuate the apoptotic effect of β-lapachone on human endothelial cells and suggest that β-lapachone may thus have potential as an anti-angiogenic drug.

In this sense, Park *et al.*, (2014) affirm that angiogenesis in the retina is a major cause of vision loss in retinopathy of prematurity (ROP), diabetic reti-nopathy and age-related macular degeneration (AMD) in each age group. Among angiogenesis-related blindness, ROP occurs through the partial regres-sion of pre-existing vessels by vaso-obliteration followed by the pathological angiogenesis in developing retinal vasculature with relative hypoxia. Thus, Park *et al.*, (2014) demonstrated that β-lapachone effectively reduced retinal ne-ovascularization in OIR without toxicity. The anti-angiogenic effect of β-lapachone was related to HIF-1α degradation and subsequent attenuation of VEGF expression. Importantly, β-lapachone–targeting HIF-1α could inhibit pathological retinal neovascularization in OIR without disturbing physiological retinal angiogenesis in the development. The results from these authors suggest that β-lapachone–targeting HIF-1α has a therapeutic potential as an anti-angiogenic drug for the ischaemic retinopathy.

Currently we have observed that the low solubility and non-specific dis-tribution of β-lapachone limit its suitability for clinical assays. However, Seoane *et al.*, (2013) formulated β-lapachone in an optimal random methylated-b-cyclodextrin/ poloxamer 407 mixture (i.e., β-lapachone ternary system) and, using human breast adenocarcinoma MCF-7 cells and immunodeficient mice, performed in vitro and in vivo evaluation of its anti-tumor effects on proliferation, cell cycle, apoptosis, DNA damage, and tumor growth. According to authors, this ternary system is fluid at room temperature, gels over 29ºC, and provides a significant amount of drug, thus facilitating intratumoral delivery, in situ gelation, and the formation of a depot for time-release. Administration of β-lapachone ternary system to MCF-7 cells induces an increase in apoptosis and DNA damage, while producing no changes in cell cycle. Moreover, in a mouse xenograft tumor model, intratumoral injection of the system significantly reduces tumor volume, while increasing apoptosis and DNA damage without visible toxicity to liver or kidney. These anti-tumoral effects and lack of visible toxicity make this system a promising new therapeutic agent for breast cancer treatment.

Hussain *et al.*, (2007) reported that β-lapachone inhibited the viability of human bladder carcinoma T24 and human prostate carcinoma DU145 cells by inducing apoptosis, which could be proved by formation of apoptotic bodies and DNA fragmentation. This investigation demonstrated that β-lapachone may be further studied as a promising agent for treatment of bladder cancer as well as providing important new insights into the possible molecular mechanisms of the anti-cancer activity of β-lapachone.

Small molecules that interfere with the indoleamine 2,3-dioxygenase 1 (IDO1) enzyme have demonstrated compelling anti-tumor properties in pre-clinical models and two such agents are currently being evaluated in clinical trials. The IDO1 enzyme is not a conventional cancer target in that it does not directly contribute to tumor cell growth or survival, but rather acts as an immune modulator involved in protecting tumors from immune-based destruction (Liu *et al.*, 2009). Recently, however, the preponderance of evidence primarily associates the pharmacological activity of β-lapachone with metabolic bioactivation by NQO1, which is expressed at elevated levels in a variety of human cancers. While NQO1 detoxifies many quinones, reduction of β-lapachone by NQO1 results in the production of reactive oxygen species that generate single-strand breaks in the DNA. This, in turn, induces hyperactivation of the DNA damage sensor poly(ADP-ribose) polymerase-1, which rapidly depletes NAD+ stores causing tumor cells to undergo a unique, caspase-independent programmed necrotic cell death (Flick *et al.*, 2013). In their studies Flick *et al.*, (2013) identified IDO1 inhibitory activity as a previously unrecognized attribute of the clinical candidate β-lapachone. Enzyme kinetics-based analysis of β-lapachone indicated an uncompetitive mode of inhibition, while computational modeling (Figure 6) predicted binding within the IDO1 active site consistent with other naphthoquinone derivatives. Inhibition of IDO1 has previously been shown to breach the pathogenic tolerization that constrains the immune system from being able to mount an effective anti-tumor response. According to authors, the finding that β-lapachone has IDO1 inhibitory activity adds a new dimension to its potential utility as an anti-cancer agent distinct from its cytotoxic properties, and suggests that a synergistic benefit can be achieved from its combined cytotoxic and immunologic effects.

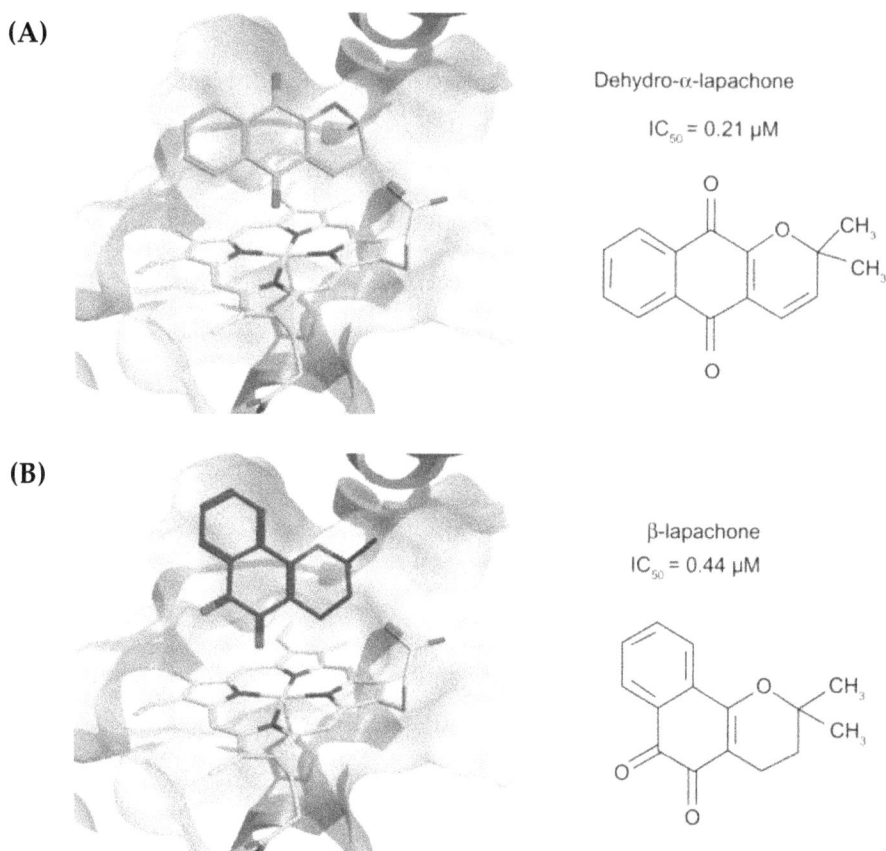

(A)

Dehydro-α-lapachone

IC_{50} = 0.21 µM

(B)

β-lapachone

IC_{50} = 0.44 µM

Figure 6: β-Lapachone is predicted to bind in the IDO1 active site. Computational modeling at the IDO1 active site and IC_{50} values for inhibition of purified human recombinant IDO1 for (A) dehydro-α-lapachone (2,2-dimethyl-2H-benzo[g]chromene-5,10-dione) and (B) β-lapachone (3,4-dihydro-2,2-dimethyl-2H-naphthol[1,2-b]pyran-5,6-dione) (Flick *et al.*, 2013).

Given all that has been explained above, we can conclude, therefore, that the b-lapachone is the most promising candidate as an effective drug in the fight against cancer.

6.3   Pterocarpanquinones

A new pterocarpanquinone (5a) was synthesized through a palladium catalyzed oxyarylation reaction and was transformed, through electrophilic substitution reaction, into derivatives 5b–d. These compounds showed to be active against human leukemic cell lines and human lung cancer cell lines. Even multidrug resistant cells were sensitive to 5a, which presented low toxicity toward peripheral blood mononuclear cells (PBMC) cells and decreased the production of TNF-a by these cells (Netto et al., 2010). The authors decided to accomplish the catalytic oxyarylation reaction of 6 using ortho-iodophenol (11). Compound 5a was not formed in condition B, but in condition C this pterocarpanquinone was obtained in 41% yield. A similar yield was obtained in the absence of PPh3 (Figure 7). Pterocarpanquinones 5b–d were obtained from 5a by electrophilic substitution, taking advantage of the great reactivity of Ering for electrophilic aromatic substitution over the A-ring, which is deactivated due to the conjugation with the carbonyl groups of the quinone moiety (Figure 8).

According to Netto et al., (2010) the pterocapanquinones 5a–d were evaluated on two human leukemic cell lines, K562 and HL-60. K562 cells, from a chronic myeloid leukemia, contains high levels of intracellular glutathione (GSH) and are resistant to oxidative stress. Cell viability greater than 90% was observed, even after treatment of these cells with H_2O_2 100 μM. In contrast, HL-60 cells, a pro-myelocytic leukemia, presents a low level of antioxidant defense and is sensible to oxidative stress. In Table 3 are presented the results obtained, showing that these new pterocarpanquinones are as active as compounds 1 and 4f, used as reference in this study. Mitomycin C was used as reference too.

Table 4 shows the IC50 obtained for 5a on Lucena-1, Raji, Jurkat and Daudi human leukemic cell lines. Lucena-1 is also resistant to oxidative stress and was slightly more resistant than K562. Jurkat, Raji and Daudi are human lymphocytic cell lines. Jurkat, a T cell leukemia, with high levels of Bcl-2 expression, was more resistant than the other leukemic cell lines. Compound 5a was very bioselective and did not show significant cytotoxicity for peripheral blood mononuclear cells (PBMC) activated by PHA ($IC_{50} > 20$ μM).

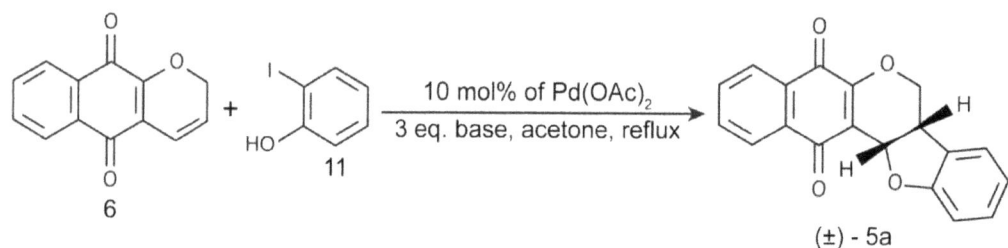

Figure 7: Synthesis of pterocarpanquinone 5a by catalytic oxy-arylation of 6.

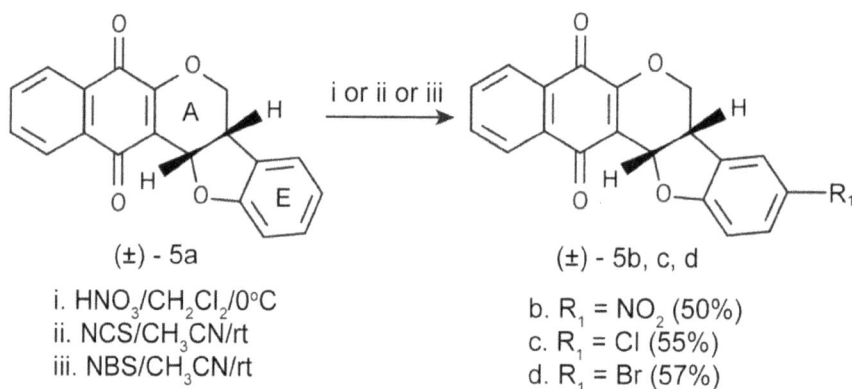

i. HNO$_3$/CH$_2$Cl$_2$/0°C
ii. NCS/CH$_3$CN/rt
iii. NBS/CH$_3$CN/rt

b. R$_1$ = NO$_2$ (50%)
c. R$_1$ = Cl (55%)
d. R$_1$ = Br (57%)

Figure 8: Synthesis of pterocarpanquinones 5b–d.

Not only leukemic cells were sensitive to quinone 5a. Table 5 shows that this quinone was very active against a small cell lung cancer cell line (GLC-4), and to a lesser extent to non-small cell lung cancer cell lines (A549 and H460).

Pterocarpanquinone 5a showed potent antineoplasic effect against leukemic cell lines, including those with the MDR phenotype, suggesting that this compound is not a substrate to ABCB1, the transporter that confers resistance to Lucena-1 cells. Although to a lesser degree, 5a was also effective against non-small lung cancer cell lines (A549 and H460) known to express different levels of the MDR transporters ABCB1 and ABCC1. The small cell lung cancer cell line GLC-4 was, however, more sensitive to 5a (Netto *et al.*, 2010).

Maia *et al.*, (2011) evaluated the anti-chronic myeloid leukemia (CML) activity and mechanisms of action of LQB-118, a pterocarpanquinone structurally related to lapachol [2-hydroxy-3-(3-methyl-2-butenyl)-1,4- naphthoquinone].

Quinone	K562	HL-60
5a	1.67	2.00
5b	3.48	0.40
5c	6.77	4.87
5d	5.70	4.87
1	2.95	2.10
3	16.04	ND
4f	2.18	ND
Mitomycin C	0.47	ND

Table 3: Antineoplasic effect of pterocarpanquinones 5a,b,d,e in K562 and HL-60 cell lines (IC$_{50}$ in μM)[3]. ND = not done.

Quinone	Lucena-1	Raji	Jurkat	Daudi	PBMC
5a	2.75	3.32	6.77	3.10	>20
Mit. C	2.75	ND	ND	0.45	4.03

Table 4: Antineoplasic effect of pterocarpanquinone 5a in leukemia cancer cell lines and PBMC (IC50 in μM)[1]. ND = not done.

Quinone	A549	H460	GLC-4
5a	11.21	12.86	5.17

Table 5: Antineoplasic effect of pterocarpanquinones 5a on lung cancer cell lines, A549, H460, and GLC-4 cell lines (IC$_{50}$, μM). ND = not done.

---

[3] Results are reported as IC50 values (concentration required to inhibit cell growth by 50%) in micromolar. Data represent the means of three independent experiments, with each concentration tested in triplicate and SD values were lower than 15%. Data from Netto *et al.*, (2010).

LQB-118 treatment resulted in an important reduction of cell viability in cell lines derived from CML, both the vincristine-sensitive K562 cell line, and the resistant K562-Lucena (a cell line overexpressing P-glycoprotein). In agreement with these results, the induction of caspase-3 activation by this compound indicated that a significant rate of apoptosis was taking place. In these cell lines, apoptosis induced by LQB-118 was accompanied by a reduction of P-glycoprotein, surviving, and XIAP expression. Moreover, this effect was not restricted to cell lines as LQB-118 produced significant apoptosis rate in cells from CML patients exhibiting multifactorial drug resistance phenotype such as P-glycoprotein, MRP1 and p53 overexpression. The data suggest that LQB-118 has a potent anti-CML activity that can overcome multifactorial drug resistance mechanisms, making this compound a promising new anti-CML agent. The authors conclude that this study provided different spectra of biologic parameters that may have implications in the understanding of the LQB-118 response in cells presenting the multifactorial multidrug resistance (MDR) phenotype. LQB-118 was capable of inhibiting Pgp, survivin and XIAP expression in an MDR CML Ph + cell line and inducing a high rate of apoptosis. In addition, this anti-apoptotic effect was observed in resistant cells from patients also exhibiting simultaneous Pgp, MRP1, and p53 overexpression (Table 6). This finding also suggests that inhibitor apoptosis proteins (IAPs) and Pgp inhibition can be a common molecular target pathway reached by LQB-118.

Bacelar *et al.*, (2013) analyzed the cell death process induced by LQB 118 in K562, a chronic myeloid leukemia cell line, and in Jurkat, a lymphoblastic acute leukemia cell line. The authors found that LQB 118 induced apoptosis in both cell lines, measuring caspase-12 and caspase-9 activation, phosphatidylserine externalization, and DNA fragmentation. The compound induced an increase in cytoplasmic calcium on both cell lines. However, the compound could only induce mitochondrial membrane depolarization on K562 cells. They showed that LQB 118 may have potential therapeutic value for leukemia, being able to overcome multiple resistance mechanisms. For them LQB 118 could induce caspase-9 activation in both cell lines. Treatment with caspase-9 inhibitor not only did not protect K562 cells from the effect of LQB 118, but also augmented the reduction of viability produced by the drug in Jurkat cells. In this work, the inhibition of cathepsin B could protect K562 and Jurkat cells from the effect of LQB 118. Therefore, it is possible that caspase-9 could be inhibiting proteases originating

Patient	Age / Gender	Sokal Score	MDR Activity	Pgp	MRP1	p53	LQB-induced Apoptosis %
1	39/F	High	-	+	+	+	81.47
2	34/F	Inter	+	+	+	+	82.08
3	29/F	High	+	+	-	+	43.07
4	22/M	High	-	+	-	-	40.32
5	17/M	Low	+	-	+	-	58.29
6	31/M	Inter	+	+	+	-	33.11
7	81/F	High	+	+	+	-	40.08
8	45/M	Low	+	+	+	-	60.20
9	31/M	Low	+	+	+	-	69.40
10	39/M	Low	-	+	+	+	67.30
11	50/F	Low	-	+	+	+	85.86
12	26/M	Inter	-	+	+	-	59.84
13	47/M	Inter	+	+	+	+	84.53

Table 6: Clinical and biological characteristics of CML patients. MDR= multidrug resistence; Pgp = P-glycoprotein; F = female; M = male; Inter = intermediate; LQB = LQB-118 3μM. Data from Maia *et al.*, (2011).

from the lysosomes, such as cathepsin B. However, according to them, further research is necessary to understand the participation of these lysosomal proteases in cell death induced by LQB 118. They found that the same compound, LQB 118, can affect tumor cells by different mechanisms, all of them important in triggering the apoptotic process. This feature allows LQB 118 to overcome different resistance mechanisms present in the tumor cell.

7 Conclusion

Tabebuia impetiginosa, also known as "ipe", is one of the various species of plants tested for chemical substances. The active chemical compound found in this plant is naphthoquinone, also known as lapachol. The *Tabebuia* genus includes about 100 species of large, flowering trees that are common to South American cities' landscapes for their beauty. Lapachol is a compound with a relatively simple chemical structure that can be easily used to form derivatives into various subgroups of compounds with increasing levels of anticancer activity both *in vitro* and *in vivo*. Although in 1974 the National Cancer Institute have ruled out the possibility of the use of lapachol in fighting cancer, due to its high toxicity, this quinone and its derivatives still remain promising drugs for cancer treatment.

Reference

American Cancer Society. Pau d'arco. Available online at http://www.cancer.org/treatment/ treatmentsandsideef- fects/complementaryandalternativemedicine/herbsvitaminsandminerals/pau-d-arco.

Asche, C. Antitumour quinines. (2005). Mini-Revew: Med Chem, 5, 449–467.

Bacelar, T. S., Silva, A. J., Costa, P. R. R. & Rumjaneka, V. M. (2013). The pterocarpanquinone LQB 118 induces apoptosis in tumor cells through the intrinsic pathway and the endoplasmic reticulum stress pathway. Anti-Cancer Drugs, 24, 73–83.

Balassiano I. T., De Paulo S. A., Henriques Silva N., Cabral M. C., & da Gloria da Costa Carvalho M. (2005). Demonstration of the lapachol as a potential drug for reducing cancer metastasis. Oncol Rep, 13(2), 329-33.

Belinsky, M., & Jaiswal, A. K. (1993). NAD(P)H:quinone oxidoreductase1 (DT-diaphorase) expression in normal and tumor tissues. Cancer Metastasis Rev, 12, 103-117.

Bolton, J. L., Trush, M. A., Penning, T. M., Dryhurst, G. & Monks, T. J. (2000). Role of quinones in toxicology. Chem Res Toxicol, 13, 135-160.

Boveris, A., Docampo, R., Turrens, J. F & Stoppani, A. O. (1978). Effect of B-Lapachone on Superoxide Anion and Hydrogen Peroxide Production in Trypanosoma cruzi. Biochem J, 175, 431-439.

Carvalho, L. H., Rocha, E. M. M., Raslan, D. S., Oliveira, A. B. & Krettli, A. V. (1998). In vitro activity of natural and synthetic naphthoquinones against erythrocytic stages of Plasmodium falciparum. Braz J Med Biol Res, 21, 485-487.

Castellanos, J. R. G., Prieto, J. M. & Heinrich, M. (2009). Red Lapacho (Tabebuia impetiginosa) – A global ethnopharmacological commodity? J. Ethnopharmacol, 121, 1-13.

Costa, W. F., Oliveira, A. B., & Nepomuceno, J. C. (2011). Lapachol as na epithelial tumor in-hibitor agent in Drosophila melanogaster heterozygote for tumor suppressor gene wts. Genet Mol Res, 10, 3236–3245.

Costa, F. N., da Silva, A. J. M., Netto, C. D., Domingos, J. L. O., Costa, P. R. R., & Leitão, G. G. (2012). Purification of a synthetic pterocarpanquinone by countercurrent chromatog-raphy. Journal of the Brazilian Chemical Society, 23, 1114-1118.

Costa, P. R. R. (2009). Natural products as starting point for the discovery of new bioactive compounds: Drug candidates with antiophidic, anticancer and antiparasitic properties. Rev Virtual Quim, 1(1), 58-66.

Craig, W. J. (1999). Health-promoting properties of common herbs. Am J Clin Nutr, 70, 491-499.

da Silva, A. J. M., Buarque, C. D., Brito, F. V., Aurelian, L., Macedo, L. F., Malkas, L. H., Hick-ey, R. J., Lopes, D. V. S., Nöel, F., Murakami, Y. L. B., Silva, N. M. V., Melo, P. A., Ca-ruso, R. R. B., Castro, N. G., & Costa. P. R. R. (2002). Synthesis and Preliminary Phar-macological Evaluation of New (±) 1,4-Naphthoquinones Structurally Related to Lapachol Bioorg. Med Chem, 10, 2731.

Dinnen, R. D. & Ebisuzaki, K. (1997). The search for novel anticancer agents: a differentiation-based assay and analysis of a folklore product. Anticancer Res, 17, 1027-1033.

Esteves-Souza A., Figueiredo, D. V., Esteves, A., Câmara, C. A., Vargas, M. D., Pinto, A. C. & Echevarria, A. (2007). Cytotoxic and DNA-topoisomerase effects of lapachol amine deriv-atives and interactions with DNA. Braz J Med Biol Res, 40, 1399-1402.

Esteves-Souza, A., Lucio, K. A., Cunha, A. S., Pinto, A. C., Lima, E. L. S., Camara, C. A., Var-gas, M. D., & Gattass, C. R. (2008). Antitumoral activity of new polyamine naphthoqui-none conjugates. Oncology Reports, 20, 225-231.

Flick, H. E., LaLonde, J. M., Malachowski, W. P. & Muller A. J. (2013). The Tumor-Selective Cytotoxic Agent β-Lapachone is a Potent Inhibitor of IDO1. Int J Tryptophan Res, 6, 35–45.

Fonseca, S. G. C., Braga, R. M. C. & Santana, D. P. (2003). Lapachol chemistry, pharmacology and assay methods. Rev Bras Farm, 84(1), 9-16.

Galm, U., Hager, M. H., Lanen, S. G. V., Ju, J., Thorson, J. S., & Shen, B. (2005) Antitumor antibiotics: bleomycin, enediynes, and mitomycin. Chem Rev, 105, 739–758 doi:10.1021/cr030117g.

Gewirtz, D. A. (1999). *A critical evaluation of the mechanisms of action proposed for the anti-tumor effects of the anthracycline antibiotics adriamycin and daunorubicin. Biochem Pharmacol, 57, 727–741 doi:10.1016/S0006-2952(98)00307-4*

Grazziotin, J. D., Schapoval, E. E., Chaves, C. G., Gleye, J. & Henriques, A. T. (1992). *Investigation phytochemistry and analgesic of Tabebuia chrysotricha. J Ethnopharmacol, 36, 249-251.*

Hussain, H., Krohn, K., Ahmad, V. U., Miana, G. A., et al., (2007). *Lapachol: an overview. Arkivoc ii, 145-171.*

Kinghorn, A. D., Farnsworth, N. R., Soejarto, D. D., Cordell, G. A., Swanson, S. M., Pezzuto, J. M., Wani, M. C., Wall, M. E., Oberlies, N. H., Kroll, D. J., Kramer, R. A., Rose, W. C., Vite, G. D., Fairchild, C. R., Peterson, R. W. & Wild, R. (2003). *News estrategies for to discovery of plants derivates how anticancer agents. Pharm Biol, 41, 53–67.*

Kongkathip, N., Luangkamin, S., Kongkathip, B., Sangma, C., Grigg, R., Kongsaeree, P., Prabpai, S., Pradidphol, N., Piyaviriyagul, S., & Siripong, P. (2004). *Synthesis of novel rhinacanthins and related anticancer naphthoquinone esters. J Med Chem, 47, 4427-4438.*

Kumagai, Y., Tsurutani, Y., Shinyashiki, M., Homma-Takeda, S., Nakai, Y., Toshikazu, Y. & Shimojo, N. (1997). *Bioactivation of lapachol responsible for DNA scission by NADPH-cytochrome P450 reductase. Environ Toxicol Pharmacol, 3, 245-250.*

Kumagai, Y. & Shimojo, N. (2002). *Possible mechanisms for induction of oxidative stress and suppression of systemic nitric oxide production caused by exposure to environmental chemicals. Environ Health Prev Med, 7(4), 141–150.*

Kung, H. N., Weng, T. Y., Liu, Y. L., Lu, K. S., & Chau, Y. P. (2014). *Sulindac Compounds Facilitate the Cytotoxicity of b-Lapachone by Up-Regulation of NAD(P)H Quinone Oxidoreductase in Human Lung Cancer Cells. Plos One. DOI: 10.1371/ journal.pone.0088122.*

Kung, H. N., Chien, C. L., Chau, G. Y., Don, M. J. Lu, K. S., & Chau, Y. P. (2007). *Involvement of NO/cGMP signaling in the apoptotic and anti-angiogenic effects of beta-lapachone on endothelial cells in vitro.Journal Cell Physiol, 211, 522-532.*

Lambertia, M. J., Vittara, N. B. R., Silva, F. C, Ferreira, V. F., & Rivarola, V. A. (2013). *Synergistic enhancement of antitumor effect of β-Lapachone by photodynamic induction of quinone oxidoreductase (NQO1). Phytomedicine, 20(11), 1007–1012.*

Larsson, D.E., Lovborg, H., Rickardson, L., Larsson, R., Oberg, K., & Granberg, D. (2006). *Identification and evaluation of potential anti-cancer drugs on human neuroendocrine tumor cell lines. Anticancer Research, 26(6B), 4125-4129.*

Lee, J. H., Cheong, J., Park, Y. M., & Choi, Y. H. (2005). Down-regulation of cyclooxygenase-2 and telomerase activity by β-lapachone in human prostate carcinoma cells. Pharmacol Res, 51(6), 553-60.

Li, Y., Sun, X., LaMont, J. T., Pardee, A. B., & Li, C. J. (2003). Selective killing of cancer cells by beta -lapachone: direct checkpoint activation as a strategy against cancer. Proc Natl Acad Sci U S A, 100, 2674-2678.

Lima, O. G., D'Albuquerque, I. L., Gonçalves de Lima, C., Machado, & M. P. Primeiras (1956). Observações sobre a Ação Antimicrobiana do Lapachol.- Anais da Sociedade de Biologia de Pernambuco, 14(1/2), 129-135.

Liu, X., Newton, R. C., Friedman, S. M., & Scherle, P. A. (2009). Indoleamine 2,3-dioxygenase, an emerging target for anti-cancer therapy. Curr Cancer Drug Targets, 9(8), 938–52.

Maia, R. C., Vasconcelos, F. C., Bacelar, T. D., Salustiano, E. J., da Silva, L. F. R., Pereira, D. L., Moellman-Coelho, A., Netto, C. D., da Silva, A. J., Rumjanek, V. M., & Costa, P. R. R. (2011). LQB-118, a pterocarpanquinone structurally related to lapachol [2-hydroxy-3-(3-methyl-2-butenyl)-1,4-naphthoquinone]: a novel class of agent with high apoptotic effect in chronic myeloid leukemia cells. Investigational New Drugs, 29(6), 1143-1155.

National Cancer Institute, 2014. Available online at: http://www.cancer.gov/drugdictionary?cdrid=357565.

Netto, C. D., da Silva, A. J., Salustiano, E. J., Bacelar, T. S., Riça, I. G., Cavalcante, M. C., Rumjanek, V. M., & Costa, P. R. R. (2010). New pterocarpanquinones: synthesis, antineoplasic activity on cultured human malignant cell lines and TNF-alpha modulation in human PBMC cells. Bioorg Med Chem, 18(4), 1610-6. doi: 10.1016/j.bmc.2009.12.073.

Newman, D.J., Cragg, G. M. & Snader, K. M. (2003). Natural products as sources of new drugs over the period 1981-2002. J Nat Prod, 66, 1022–1037.

Oliveira, M. F. (2000). Contribuição ao conhecimento químico das espécies Tabebuia serratifolia Nichols e Tabebuia rosa Bertol. Tese de Doutoramento, Curso de Pós-graduação em química orgânica, Universidade Federal do Ceará, Fortaleza, Ceará, Brasil.

Park, H. J., Ahn, K. I., Ahn, S. D., Choi, E., Lee, S. W., Williams, B., Kim, E. J., Griffin, R., Bey, E. A., Bornmann,W. G.; Gao, J., Park, H. J., Boothman, D. A., & Song, C. W. (2005). Susceptibility of cancer cells to β-lapachone is enhanced by ionizing radiation. International Journal of Radiation Oncology, Biology, Physics, 61, 212–219.

Park, S. W., Kim, J. H., Kim, K.-E., Jeong, M. H., Park, H., Park, B., Suh, Y.-G., Park, W. J. & Kim, J. H. (2014), Beta-lapachone inhibits pathological retinal neovascularization in oxygen-induced retinopathy via regulation of HIF-1α. *Journal of Cellular and Molecular Medicine*. doi: 10.1111/jcmm.12235.

Rao, K. V., McBride, T. J. & Oleson, J. J. (1968). *Recognition and Evaluation of Lapachol as an Antitumor Agent. Cancer Res*, 28, 1952-1954.

Ravelo, A. G, Estévez-Braun, A., Chávez-Orellana, H., Pérez-Sacau, E. & Mesa-Siverio, D. (2004). *Recent studies on natural products as anticancer agents. Curr Top Méd Chem*, 4, 241-265.

Reinicke, K. E., Bey, E. A., Bentle, M. S., Pink, J. J., Ingalls, S. T., Hoppel, C. L., Misico, R. I., Arzac, G. M., Burton, G., Bornmann, W. G., Sutton, D., Gao, J., & Boothman, D. A. (2005). *Development of beta-lapachone prodrugs for therapy against human cancer cells with elevated NAD(P)H:quinone oxidoreductase 1 levels. Clin Cancer Res*, 11, 3055-3064.

Seoane, S., Díaz-rodríguez, P., Sendon-lago, J., Gallego, R., Pérez-fernández, R., & Landin, M. (2013). *Administration of the optimized b-Lapachone–poloxamer–cyclodextrin ternary system induces apoptosis, DNA damage and reduces tumor growth in a human breast adenocarcinoma xenograft mouse model. European Journal of Pharmaceutics and Biopharmaceutics*, 84(3), Pages 497–504.

Siegel, D. D. L., Gustafson, D. L., Dehn, J. Y., Han, P., Boonchoong, L. J., & Ross, D. (2004). *NAD(P)H:quinone oxidoreductase 1: role as a superoxide scavenger. Mol Pharmacol*, 65, 1238-1247.

Silva, M. N., Ferreira, V. F. & Souza, M. C. B. V. (2003). *An overview of the chemistry and pharmacology of naphthoquinone with emphasis on β-lapachone and derivatives. Quim Nova*, 26, 407-416.

Suffness, M. & Douros, J. (1980). *Miscellaneous natural products with antitumor activity. In: J. M. Cassady, J. D. Douros, (Eds.), Anticancer Agents Based on Natural Product Models*, Academic Press, New York, pp 474 (Chapter 14).

Taylor, L. (2005). *The Healing Power of Rainforest Herbs: A Guide to Understanding and Using Herbal Medicinals*. SquareOne publishers, 519 pgs.

Vieira, R. F & Martins, M. V. M. (1998). *Estudos etnobotânicos de espécies medicinais de uso popular no Cerrado, In: Proc. Int. Savanna Simposium, Brasilia, DF, Embrapa/CPAC*, pp 169–171.

Wolkenberg, S. E., & Boger, D. L. (2005). *Mechanisms of in situ activation for DNA-targeting antitumor agents. Chem Rev*, 102, 2477–2495 doi:10.1021/cr010046q

Woo, H. J. & Choi, Y. H. (2005). Growth inhibition of A549 human lung carcinoma cells by beta-lapachone through induction of apoptosis and inhibition of telomerase activity. Int J Oncol, 26, 1017-1023.

Zhang, G., Fang, L., Zhu, L., Zhong, Y., Wang, P. G., & Sun, D. (2006). Syntheses and biological activities of 3'-azido disaccharide analogues of daunorubicin against drug-resistant leukemia. J Med Chem, 49, 1792–1799 doi:10.1021/jm050916m

DNA Barcoding Assessment of the Genetic Diversity of Varieties of Taro, *Colocasia Esculenta* (L.) Schott in Brazil

Raquel Soares Casaes Nunes[1], Eduardo Mere Del Aguila[1],
Vânia Margaret Flosi Paschoalin[1], Joab Trajano da Silva[1]

1 Introduction

Taro, *Colocasia esculenta* (L.) Schott, generally called "inhame" (yam), is a plant with an unconventional tuber, an herbaceous species of the family Araceae (class Liliatae). Taro grows in tropical and subtropical climates and can thrive in adverse conditions, such as poor soils, excess soil water, shady conditions or climatic stress (Xu *et al.*, 2001; Zárate *et al.*, 2009).

Taro originated in southern Asia and most of the islands between Asia and Australia, known as the Indo-Malayan region; it was then distributed to east and southeast Asia, the Pacific islands, Madagascar and Africa, and then brought to tropical America (Ivancic & Lebot, 2000). Colocasia is usually cultivated by family agriculturists in traditional populations for their own consumption and for exchange for other products (Yalu *et al.*, 2009; Singh *et al.*, 2007). In Brazil, taro has a limited distribution in the midwestern and southeastern regions. It is a significant component of the food consumed by traditional communities; however, the crops are not organized into formal production chains as are conventional vegetables such as potatoes, tomatoes, cabbage and lettuce (Diegues, 2001; Madeira *et al.*, 2008; FAO, 2010).

Encouraging the cultivation and propagation of cultivars can stimulate the conservation of a species, thus preventing genetic losses by flowering or

[1] Instituto de Química, Universidade Federal do Rio de Janeiro, Rio de Janeiro, Brazil

microbial deterioration. Genetic studies can identify morphological varieties of taro, classifying them as cultivars based on the genetic differences among them, and may collect information about genes that confer nutritional value. The Brazilian cultivars of taro arise from vegetative mutations; according to Puiatti *et al.* (2003), the best-known cultivars of *Colocasia* are the "Macaquinho", "Chinês", "Japonês" and "São Bento" (Figure 1).

Figure 1: Tubers of *Colocasia esculenta* preserved in the germ-plasm bank of the Instituto Capixaba de Pesquisa, Assistência Técnica e Extensão Rural (INCAPER) located in Espírito Santo state, Brazil. (A1) Aerial part T38 (Chinês) plant size 60 cm from midwestern Brazil; (B1) Aerial part T37 (Macaquinho) plant size 49,7 cm from midwest; (C1) Aerial part T42 (São Bento) plant size 80 cm from, southeast. Rhizomes: (A2) T38 (Chinês) rhizomes size 5cm from midwest: (B2) T37 (Macaquinho) rhizomes size 5cm from Midwest; (C2) T42 (São Bento) rhizomes size 5cm from southeast. (Zárate *et al.*,2009). Source: INCAPER /Espírito Santo, 2011.

In Brazil, the Chinês, Japonês and Macaquinho varieties are grown in the southeast (Minas Gerais state) and midwest (Mato Grosso do Sul state) (Gondim *et al.*, 2007). The cultivars "Branco", Chinês and Japonês can be found in Rio de Janeiro state, and others, "Rosa" and "Roxo", are also common in this southeast region (Oliveira *et al.*, 2004; Oliveira *et al.*, 2008; Silva *et al.*, 2006). The state of Espírito Santo, also located in the southeast, is a traditional producer of taro, and the predominating clones found there are Chinês and São Bento (Carmo & Puiatti, 2004).

The genetic-characterization search for a plant DNA barcode has focused on genes or sequences of the chloroplast genome, and several candidates such as *matK, rbcL* and *psbA-trnH* have been proposed (Hollingsworth *et al.*, 2009; Sun *et al.*, 2012; Kress *et al.*, 2005; Newmaster *et al.*, 2006; Yoo *et al.*, 2006). The *rbcL* gene encodes the enzyme ribulose-1,5-bisphosphate carboxylase oxidase (Cuénoud *et al.*, 2002), and the *matK* gene encodes the maturase K enzyme (Kress *et al.*, 2005) and *psbA-trnH*, which is an intergenic spacer sequence (Xu *et al.*, 2000). The genetic resources are conserved using *ex situ* conservation in germ-plasm banks (Nass *et al.*, 2001; Carmo & Puiatti, 2004; Mace *et al.*, 2006; Singh *et al.*, 2011), a procedure that has been adopted by several plant breeding programs (TANSAO, 2000; TAROGEN, 2001). In Brazil, taro cultivars are maintained in the germ-plasm bank of the Instituto Capixaba de Pesquisa, Assistência Técnica e Extensão Rural (INCAPER), located in the city of Vitória, the state capital of Espírito Santo.

The use of barcode loci to explore intra-species variation (Sun *et al.*, 2012; Singh *et al.*, 2012) has been underexplored. In this study, the sequence analysis of exons of the plastid genes *rbcL* and *matK* loci and the *pbsA-trnH* intergenic spacer were used singly and in combination to assess the relationships among taro cultivars held in a Brazilian germ-plasm bank. We also addressed the question of whether these three loci could correctly reflect the geographic origin of cultivars.

2 Materials and Methods

2.1 Plant Material

Seven taro cultivars were obtained from the INCAPER collection. According to the morphological characteristics, they were named and classified as cultivars

T37 (Macaquinho), T38 (Chinês), T39 (Japonês), T40 (Chinês Regional), T41 (Cem em Um), T42 (São Bento) and T43 (Branco). T37, T38, T39, T41 and T43 were collected in midwestern Brazil, and T40 and T42 were collected in southeastern Brazil. Rhizome and leaf samples of each cultivar were collected, dried and kept at 70°C for 4-5 days.

2.2   DNA Preparation

Genomic DNA was extracted using the DNeasy Plant Mini kit (Quiagen Inc., Germany) and quantified using the Qubit™ fluorimeter (Invitrogen/Life Technologies Inc., USA) and the Qubit dsDNA HS Assay Kits (Invitrogen /Life Technologies Inc., USA). Aliquots of 5 µl were applied to 2% agarose gel, and electrophoresis was carried out in 1X TAE buffer at 100 V and 200 mA for 90 min, followed by staining in a solution of 100 mL of GelRed (BioAmerica Inc., USA). DNA fragments were recorded digitally by using the Bio-Imaging System (BioAmerica Inc., USA).

2.3   DNA Sequence Amplification

Amplifications were performed with the use of a thermocycler (MyCycler™, Bio-Rad, Berkeley, USA). Three primer sets targeting for *matK* (www.barcoding.si.edu) and *rbcL* (Cuénoud *et al.*, 2002) loci and the *psbA-trnH* intergenic spacer (Kress *et al.*, 2005) were used. Amplicons obtained by using the *rbcL* primers (forward: ATGTCACCACAAACAGAGACTAAAGC, reverse GTAAAATCAAGTCCACCG) were 654-1434 bp, and amplicons obtained by using the *matK* primers (forward: CGATCTATTCATTCAATATTTC, reverse: TCTAGCACACGAAAGTCGAAT) were 1746 bp. Amplicons obtained by using the *psbA-trnH* spacer primers (forward: TGATCCACTTGGCTACATCCGCC, reverse: GCTAACCTTGGTATGGAA) ranged from 318 to 820 bp. PCR was performed using 25 ng of genomic DNA, 0.2 nM of each primer, 0.1 mM of each dNTP, 2.5 mM $MgCl_2$ and 0.5 U of Taq DNA polymerase (Promega) in a final volume of 25 µL. PCR was performed under the following conditions: 96°C for 2 min, followed by 35 cycles at 94°C for 50 s, 55°C for 50 s and 72°C for 1 min, and a final extension at 72°C for 5 min. PCR products and standard DNA fragments of low molecular weight (GeneRuler™ Low Range Ladder, Fermentas Inc., Canada) were fractionated by electrophoresis on 2% agarose gel (Fermentas Inc., Canada) under the conditions described above.

2.4   DNA Sequencing and Data Analysis

PCR products were purified using the PCR Product Purification GFX™ PCR DNA and Gel Band Purification Kit (GE Healthcare Life Science Inc., USA) and quantified as described above. DNA sequencing was performed with the use of a BigDye® Terminator v3.1 kit (Applied Biosystems Inc., USA) and an ABI-3730 automatic sequencer (Applied Biosystems Inc., USA). Sequence similarities were searched by BLASTn (Zhang *et al.*, 2000), in GenBank, available at the National   Center   for   Biotechnology   Information   (http://www. ncbi.nlm.nih.gov).

2.5   Phylogenetic Analysis

Phylogenetic relationships among the cultivars were analyzed from the alignment of sequences in Clustal X 2.0 software (Larkin *et al.*, 2007) for constructing the phylogenetic trees, using the Mega 5.02 software and UPGMA methods (Sneath & Sokal, 1973).

Phylogenetic analysis of the three-region combined (*rbcL;matK and psaB-trnH*) was performed in Clustal X 2.0 software using the Mega 6.0 software

3   Results

3.1   Taxonomy of *Colocasia esculenta* Cultivars

Sequencing analysis of the *rbcL* locus identified the seven cultivars as *C. esculenta* with 96-100% homology. However, it was not possible to distinguish among *C. esculenta, Steudnera colocasiifolia* and *Remusatia yunnanensis* (Table 1; Annex 1), since five cultivars, T37, T38, T40 and T42, showed 98 to 99% homology to *C. esculenta, 98 to 99 % to S. colocasiifolia and 99% to Remusatia yunnanensis*. One cultivar, T41 was identified as *C. esculenta,* and also as *Colocasia menglaensis* with 99 % homology, only the T43 was identified as single *C. esculenta.*

By sequencing the *matK* locus, the seven cultivars were classified as *C. esculenta,* with 97 to 100% similarity. Five cultivars T37,T38, T39, T40, T 43 showed 97- 99 % similarity to *Colocasia menglaensis* (Table 2, Annex 2), one cultivar T41 showed 98% similarity to *Remusatia vivipara,* one cultivar, T42 showed 98% similarity to *Remusatia pumila* (Table 2). However, like the

Sample Number	GenBank Accession Number	Species Identified	Similarity (%)
T37	JQ933275.1	*Colocasia esculenta*	99
	JF828101.1	*Steudnera colocasiifolia*	99
	JF828096.1	*Remusatia yunnanensis*	99
T38	JQ933275	*Colocasia esculenta*	99
	JF828101.1	*Steudnera colocasiifolia*	99
	JF828096.1	*Remusatia yunnanensis*	99
T39	JQ933275	*Colocasia esculenta*	99
	JF828101.1	*Steudnera colocasiifolia*	99
T40	JQ933275	*Colocasia esculenta*	98
	JF828101.1	*Steudnera colocasiifolia*	98
T41	JQ933275	*Colocasia esculenta*	99
	JQ238894.1	*Colocasia menglaensis*	99
T42	JQ933275	*Colocasia esculenta*	98
	JF828101.1	*Steudnera colocasiifolia*	98
	JF828096.1	*Remusatia yunnanensis*	98
T43	JN105690.1	*Colocasia esculenta*	99

Table 1: Identification of taro cultivars by comparison with the 700 bp amplicons from the *rbcL* loci with sequences in GenBank, using the BLASTn algorithm.

sequencing the *rbcL* locus, it was impossible to distinguish them from *Colocasia menglaensis*.

Nucleotide sequencing of the *pbsA-trnH* intergenic spacer demonstrated a similarity of 99% between the seven cultivars and *C. esculenta* and it is was able to distinguish this species from others *Colocasia sp.* in six of these cultivars: T37, T38 and T43 cultivars showed 94-96% similarity to *Colocasia antiquorum*, T39 cultivar showed 87% similarity to *Colocasia yuannanensis*, T40, T42 and T43 cultivar showed 97-98% similarity to *Colocasia lihengiae* and T42 cultivar showed 88% similarity *Colocasia heterocroma)* (Table 3, Annex 3). The sequence analysis

Sample Number	GenBank Accession Number	Species Identified	Similarity ID (%)
T37	JN105690.1	*Colocasia esculenta*	99
	JQ238892.1	*Colocasia menglaensis*	99
T38	JN105690.1	*Colocasia esculenta*	100
	JQ238892.1	*Colocasia menglaensis*	100
T39	JN105690.1	*Colocasia esculenta*	99
	JQ238894.1	*Colocasia menglaensis*	99
T40	JN105690.1	*Colocasia esculenta*	99
	JQ238894.1	*Colocasia menglaensis*	99
T41	JN105690.1	*Colocasia esculenta*	99
	EU886584.1	*Remusatia vivipara*	98
T42	JN105690.1	*Colocasia esculenta*	99
	JQ238896.1	*Remusatia pumila*	98
T43	JN105690.1	*Colocasia esculenta*	97
	JQ238894.1	*Colocasia menglaensis*	97

Table 2: Identification of taro cultivars by comparison of the 900 bp amplicons from the *matK* loci with sequences in GenBank, using the BLASTn algorithm.

of T41 cultivar did not differentiated *C. esculenta* voucher SB Davis 1225, an invasive species found in Florida, USA (accession GU 135448.1), from the ordinary accession number of *C. esculenta* (JN105690.1) (Table 3, Annex 3).

3.2   Phylogeny of *Colocasia Esculenta* Cultivars

Alignment of DNA sequences from *rbcL*, *matK* and *psbA-TrnH* in combination grouped the cultivars into three clusters. The phylogenetic tree was constructed using the locus *rbcL* as the root. A bootstrap value of 1.0 was obtained between the *rbcL and matK* loci, and a bootstrap value of 0.5 was found for the *matK* locus and the *psbA-TrnH* intergenic sequence. In the alignment of *rbcL* and *matK* loci,

Sample Number	GenBank Accession Number	Species Identified	Similarity (%)
T37	JN105690.1	*Colocasia esculenta*	99
	JF828132.1	*Colocasia antiquorum*	96
T38	JN105690.1	*Colocasia esculenta*	99
	JF828132.1	*Colocasia antiquorum*	94
T39	JN105690.1	*Colocasia esculenta*	99
	JF828131.1	*Colocasia yunnanensis*	87
T40	JN105690.1	*Colocasia esculenta*	99
	JF828137.1	*Colocasia lihengiae*	97
T41	JN105690.1	*Colocasia esculenta*	99
T42	GU135448.1	*Colocasia esculenta*	99
	JN105690.1	*Colocasia esculenta*	98
	JF828137.1	*Colocasia lihengiea*	97
	JF828138.1	*Colocasia heterocroma*	88
T43	JN105690.1	*Colocasia esculenta*	99
	JF828137.1	*Colocasia lihengiea*	98
	JF828132.1	*Colocasia antiquorum*	95

Table 3: Identification of taro cultivars by comparison of the 550 bp amplicons from the *pbsA-trnH* intergenic spacer with sequences in GenBank, using the BLASTn algorithm.

there was no bootstrap-value distance among the subclusters containing the cultivars T37, T38, T39, T40, T41, T42, T43 in the *matK* locus; however, the T43 cultivar showed a bootstrap value of 1.0 for the *rbcL* locus. The phylogenetic tree obtained from the *rbcL* loci indicated a close genetic relationship among the cultivars. All the cultivars were grouped in the same cluster, with the exception of the T43 cultivar (Figure 3A, Annex 4).

The alignment of the *psbA-TrnH* spacer sequence showed a bootstrap value of 0.1 for the cultivars T38, T40 and T42 (Figure 2). Alignment of DNA sequences from the *matK* loci (Figure 2; 3B; Annex 5) discriminated the cultivars

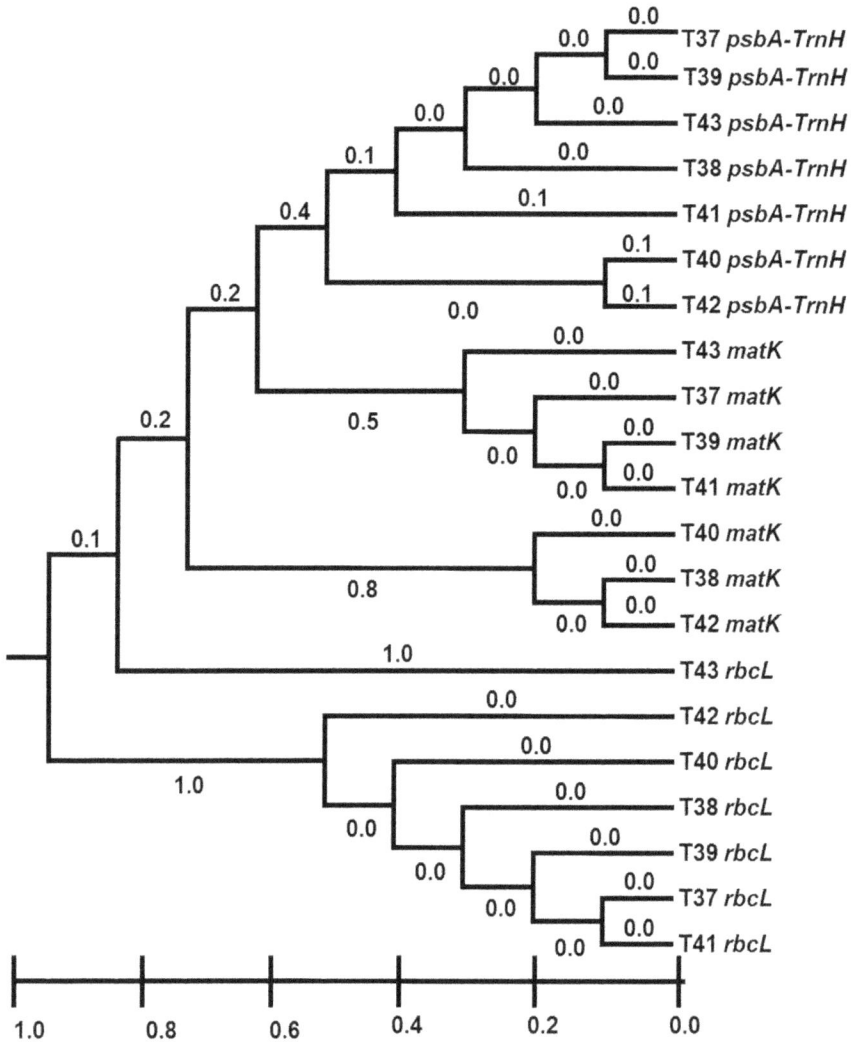

Figure 2. Phylogenetic trees generated from multiple alignment of the sequences *rbcL* loci, *matK* loci and *pbsA- trnH* intergenic spacer. The taro cultivar collection comprises the set from INCAPER where cultivars T37, T38, T39, T41 and T43 were collected in midwestern Brazil, whereas T40 and T42 were collected in southeastern Brazil.

(A)

(B)

(C)

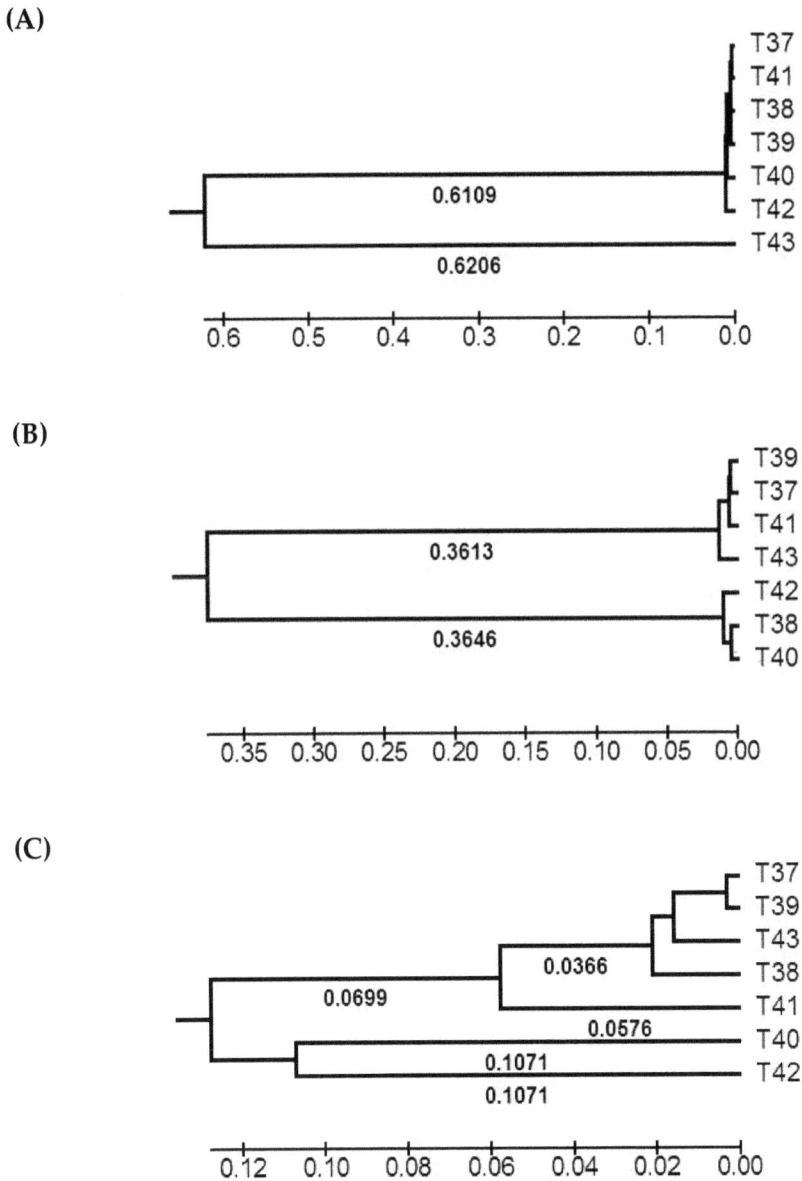

Figure 3. Phylogenetic trees generated from the alignment of the sequences *rbcL* loci (**A**), *matK* loci (**B**) and *pbsA-trnH* intergenic spacer (**C**) in Clustal X 2.0 software using the Mega 5.02 and UPGMA methods. The taro cultivar collection comprises the set from INCAPER, where cultivars T37, T38, T39, T41 and T43 were collected in midwestern Brazil, whereas T40 and T42 were collected in southeastern Brazil.

into two clusters. The first cluster grouped the cultivars T37, T39, T41 and T43, all of them collected in midwestern Brazil, specifically in Mato Grosso do Sul; while the second cluster grouped two cultivars that were originally collected in southeastern Brazil, in Espírito Santo. The cluster set comprised by cultivars T38, T40 and T42 is consistent with the origin of cultivar T42 (São Bento), which was asexually propagated from T40 (Chinês) in southeastern Brazil, itself derived from the original T38 cultivar from midwestern Brazil. In contrast, the relationship among the cultivars grouped in the first cluster cannot be determined, since the origins of T37, T39, T41 and T43, collected in midwestern Brazil, are not documented.

The divergence among varieties was demonstrated when the sequence alignment of the *psbA-trnH* intergenic spacer was considered (Figure 3C, Annex 6). T40 and T42, both from southeastern Brazil, were grouped into the same cluster, correctly reflecting their origin by vegetative propagation (Zárate & Vieira, 2006). However, T38 was grouped in another main cluster, in a subcluster comprising the T37, T39 and T43 cultivars from midwestern Brazil.

4   Discussion

The three nucleotide sequences from plastid genes and the intergene region (*rbcL, matK* and *psbA-trnH*), used singly or in combination, allowed the identification of the Brazilian taro collection from the INCAPER germ-plasm bank as *C. esculenta*. However, the *rbcL* sequences were not able to discriminate the taro cultivars from the genera *Steudnera* and *Remusatia*.

The taxonomic analyses of Brazilian taro cultivars based on the *rbcL* locus sequence were not sufficiently accurate to classify them as *C. esculenta* (JQ933275.1and JN105690.1) and they were also identified as *S. colocasiifolia*. Sequencing analysis of the *matK* locus was not able to discriminated the culti-vars as members of a single genus: on the contrary, they were classified as *C. esculenta, Remusatia vivipara, and Remusatia pumila*. Li *et al.* (2012) reinforced the genetic similarities among *Remusatia* spp., *Steudnera colocasiifolia* and *C. esculenta* within the Colocasia clade by analyzing sequence data from three plastid regions (the rbcL gene, the trnL-trnF intergenic spacer, and the rps16 intron).

On the other hand, the analysis of the *pbsA-trnH* intergenic spacer sequences identified the cultivars which *C. esculenta* discriminate into of others species of genus *Colocasia sp.* as *Colocasia Antiquorum, Colocasia lihengiae and Colocasia heterocroma,* on the contrary of sequences analysis of cultivars T37, T38, T39, T40 , T43 with matK locus that showed high similarity with species belonging the this genus Colocasia *sp.(Colocasia menglaensis).*

The phylogenetic relationships within the family Araceae have been addressed by several researchers. According to Mayo *et al.* (1997), Araceae includes 105 genera and 3300 species occurring on all continents except Antarctica. The relationships among the family members were reconstructed on the basis of similarities and differences in morphological attributes. In 2008, Cabrera *et al.* evaluated the familial, subfamilial and tribal monophyly and relationships of aroids through a phylogenetic analyses of five regions of coding (*rbcL, matK*) and noncoding plastid DNA (partial *trnK* intron, *trnL* intron, *trnL – trnF* spacer). They classified several genera within the tribe Colocaseae; Colocasia, Steudnera *and* Remusatia *were allocated to the same clade with 100% bootstrap value. In recent studies, it was* studied non coding sequences in the taro samples (*C. esculenta*) and *suggested that some species of* C. *esculenta and Colocasia may be closer than is currently thought* (Nauheimer *et al., 2012 and Ahmed et al., 2013).* Zuo, 2011 recommended that two-locus combinations rbcL-trnH-psbA can be used for distinguishing and identifying Colocasieae species.

In this study, the plastid sequences – rbcl, matK and pbsA-trnH – allowed us to analyze the genetic relationships among the Brazilian taro varieties.

The *rbcL* gene, which encodes the largest subunit of the enzyme ribulose-1,5-bisphosphate carboxylase oxidase, Rubisco (Cuénoud *et al.,* 2002), as mentioned before, evolves slowly, giving accurate results for higher taxonomic levels such as family. However, it is not considered very useful for phylogenetic studies at the species level (Chase *et al.,* 1993), since low-frequency mutations were observed for this gene. Indeed, the *rbcl* locus analysis of the seven cultivars did not indicate a divergence within the cultivars, grouping six of them together, and discriminating only T43.

The chloroplast maturase K gene (*matK*) is one of the most variable coding genes of angiosperms, and has been suggested for use as a "barcode" for land plants. However, *matK* exhibits low rates of both amplification and sequencing due to the low universality of the currently available primers and mononucleotide repeats (Yu *et al.*, 2011). The *matK* sequence analysis of the seven cultivars grouped them correctly according to their genetic similarities, even with the incomplete sequencing of the fragment (900 bp nucleotides from 1746 bp).

On the other hand, the plastid *psbA-trnH* intergenic spacer regions were completely sequenced, and 550 from a total of 562 pairs of nucleotides were obtained. This could be considered a short sequence to establish genotype relationships, but it is the most commonly sequenced locus used in plant phylogenetic investigations at the species level, and it shows high levels of interspecific divergence (Kress *et al.*, 2005). This non-coding sequence is currently used for establishing the lower taxonomic levels such as genera and species, since they tend to accumulate mutations more rapidly (Kress & Erickson, 2007). The relationships among a diverse set of genera from seven plant families were investigated using sequences of three chloroplast regions, *psbA-trnH*, *p136-rpf8* and *trnL-F*. The percentage of sequence divergence of the *psbA-trnH* intergenic spacer was three times higher than for the other two regions (1.24% in *psbA-trnH* compared to 0.44% in both *rp136-rpf8* and *trnL-F* (Kress *et al.*, 2005; Cuénoud *et al.*, 2002). In this study, the non-coding region *psbA-trnH* showed greater intra- and interspecific divergence than the coding regions *matK* and *rbcL* (Whitlock *et al.*, 2010).

The analysis of *pbsA-trnH* sequence from the seven cultivars showed genetic variability among them. The cultivars were grouped in two separate clades, gathering them correctly into cultivars from the Midwest and Southeast. Perhaps the differences observed among the genetic relationships of the taro cultivars when considering the *matK* and *psbA-trnH* nucleotide sequence markers might be attributed to the incomplete sequencing of the *matK* locus, and could require a more accurate study.

To improve the intraspecies-specific analysis, a combination of barcode loci matK+rbcL+psbA-trnH *was performed, as suggested previously by* Hollingsworth *et al.* (2009) who combined *matK+rbcL*, and by Kress & Erickson (2007), who combined the trnH-psbA spacer with either rbcL-a, rpoB2, or rpoC1. *Ahmed et al., 2013 performed a multiple sequence alignments for the intraspecific comparison of* C. esculenta at 30 loci and microsatellite polymorphisms at loci in order to identify mutational hotspots and loci suitable for *population genetic phylogenetic and phylogeographical studies*

Analysis of the phylogenetic trees obtained using singly or combined loci and intergenic spacers confirmed the close genetic relationship among the cultivars. In the alignment of DNA sequences of matK+rbcL+psbA-trnH with the three loci, the cultivars amplifying with the *rbcL* and *matK* loci obtained bootstrap values of 0.0, and then were considered identical. However, considering the alignment of the *psbA-TrnH* intergenic spacer, there is a genetic distance between the cultivars T40 and T42, and the subcluster formed by T37, T38, T39, T41 and T43, considering the bootstrap of 0.1. The divergence between T38 and T40/T42 is not clearly understood, since T40 and T42 were asexually propagated, but it is possible that de novo mutations may have been acquired.

Nunes *et al.* (2011) assessed the genetic diversity of the same collection of taro cultivars, using a set of seven microsatellite markers: Xuqtem55, Xuqtem73, Xuqtem84, Xuqtem88, Xuqtem91, Xuqtem97 and Xuqtem110 loci. The polymorphic analysis contributed to discriminate the *C. esculenta* cultivars: the Xuqtem88 locus was considered the best marker, and the results obtained are similar to those obtained with the non-codifying *psbA-trnH* intergenic spacer concerning the closeness among the southeastern cultivars T40 (Chinês) and T42 (São Bento) and the higher diversity among the Brazilian midwestern cultivars.

In analyzing the Xuqtem110 locus, similar results were obtained with *matK* sequences concerning the parental origin of the T40 (Chinês) and T42 (São Bento) cultivars. Taken together, both microsatellite markers (Mace & Godwin, 2002; Mace *et al.*, 2006; Nunes *et al.*, 2011) and universal primers coding for the maturase K gene and targeting for non-coding regions such as *psbA-trnH* can contribute to the study of genetic diversity of *C. esculenta* (Figure 2, B and C).

The genetic relationships among cultivars seem to be related to their productivity. Biomass production is desirable, and taro cultivars with large and uniform rhizomes command the best prices in a demanding market (Zárate *et al.*, 2006; 2009). A previous study established that the productivity of taro is influenced by differences in planting practices and not by the genetic features of the different cultivars (Zárate & Yamaguti, 1994). This statement makes it impossible to select the best cultivar with appropriate characteristics for planting. Additional studies of systematic cropping of taro cultivars are necessary to establish the correlation between genetic analyses and the productivity, nutritional and sensory characteristics of Brazilian cultivars. However, the genetic similarities among taro cultivars described in this study conform to the known productivities of the different cultivars. Cultivars T37 T38, and T39 are superior in commercial rhizome production to cultivars T43 and T41 (Carmo & Ferrão, 2000; Zárate & Vieira, 2004; Zárate *et al.*, 2006). The new T42 cultivar, derived from the original T40, is widely consumed in Espírito Santo (Carmo & Borel, 2002; Carmo & Puiatti, 2004). Regarding the plant productivity, T37 and T38, grouped in the same cluster by the *pbsA-trnH* intergenic sequence analysis, were considered the most productive corm and rhizomes, and also show similar nutritional characteristics (Zárate *et al.*, 2006; Zárate *et al.*, 2012).

5   Conclusions

This study aimed to contribute to the knowledge of the taxonomy and phylogeny of Brazilian cultivars of *Colocasia esculenta*.

Analysis of the chloroplast genome sequences such as *rbcL* and *pbsA-trnH* can be a valuable tool in establishing the phylogenetic analysis and variability of taro cultivars grown in Brazil.

The genetic evaluation of these cultivars will lead not only to a better understanding of the characteristics of the taxonomic varieties, but can also provide information about the genetic basis of differences in productivity and in the organoleptic characteristics of this crop.

Acknowledgments

This study was supported by FAPERJ (Fundação Carlos Chagas Filho de Amparo à Pesquisa do Estado do Rio de Janeiro). The authors are grateful to the Platform Sequencing of the Instituto Oswaldo Cruz (IOC-FIOCRUZ) for making a sequencer available. The authors thank Dr Carlos Alberto Simões do Carmo for providing the cultivars of *C. esculenta* from NCAPER (Instituto Capixaba de Pesquisa e Assistência Técnica e Extensão Rural).

References

Ahmed, I., Matthews, P. J., Biggs, P. J. M., Naeem, M., McLenachan, P. A., & Lockhart, P. J. (2013). Identification of chloroplast genome loci suitable for high-resolution phylogeographic studies of Colocasia esculenta (L.) Schott (Araceae) and closely related taxa. Molecular Ecology Resources, 13, 929–937.

Cabrera, L. I., Salazar, G. A., Chase, M. W., Mayo, S. J., Bogner, J., & Davila, P. (2008). Phylogenetic relationships of aroids and duckweeds (Araceae) inferred from coding and noncoding plastid DNA. American Journal of Botany, 95, 1153–1165.

Carmo, C. A. S. & Borel, R. M. A. (2002). Situação das culturas do taro e do inhame no Estado do Espírito Santo. In: Simpósio Nacional Sobre as Culturas do Inhame e do Taro, II., 2002, João Pessoa - PB. Anais. João Pessoa, Paraíba: EMEPA-PB, 197-212.

Carmo, C. A. S. & Ferrão, M. A. G. (2000). Comportamento de clones de inhame na região Centro-Serrana do Estado do Espírito Santo. In: Congresso Brasileiro de Olericultura, 40. Rio de Janeiro/RJ. Anais. Horticultura Brasileira, 18, 591-593. Suplemento.

Carmo, C. A. S. & Puiatti, M. (2004). Avaliação de clones de taro para cultivo no Estado do Espírito Santo. Horticultura Brasileira, 22, 430. Suplemento 2.

CBOL. A Consortium for the Barcode of Life. Available at www.barcoding.si.edu, accessed on January 10, 2011.

Chase, M. W., Salamin, N., Wilkinson, M., Dunwell, J. M., Kesanakurthi, R. P., Haidar, N., & Savolainen, V. (2005). Land plants and DNA barcodes: short-term and long-term goals. Philosophical Transactions of the Royal Society, Series B, Biological Sciences, 360, 1889–1895.

Cuénoud, P. V., Savolainen, W., Chatrou, L. W.., Powell, M., Grayer, R. J., & Chase, M. W. (2002). Molecular phylogenetics of Caryophyllales based on nuclear 18S rDNA and plastid rbcL, atpB and matK DNA sequences. American Journal of Botany, 89, 132–144.

Cusimano, N., Bogner, J., Mayo, S. J., Boyce, P. C., Wong, S. Y., Hesse, M., Hetterscheidt, W. L. A., Keating, R. C., & French, J. C. (2011). Relationships within the Araceae: comparisons of morphological patterns with molecular phylogenies. American Journal of Botany, 98, 654–668.

FAO Faostat: FAO Statistical Database: Agricultural Production of Primary Crops Available at http://www.ibb.unesp.br/servicos/publicacoes/rbpm/pdf_v7_n1_2005/artigo_11_v7_n1.p df, accessed on: October 22, 2010.

Gondim, A. R. O., Puiatti, M., Cecon, P. R., & Finger, F. L.(2007). *Crescimento, partição de fotoassimilados e produção de rizomas em taro cultivado sob sombreamento artificial. Horticultura Brasileira, 2(3), 418-428.*

Hollingsworth, P. M., Forrest, L. L., Spouge, J. L., Hajibabaei, M., Ratnasingham, S., van der Bank, M., Chase, M. W., Cowan, R. S., Erickson, D. L., Fazekas, A. J.. Graham, S. W., James, K. E., Kim, K.-J., Kress, W. J., Schneider, H., van Alphenstahl, J., Barrett, S. C. H., van den Berg, C.; Bogarin, D.; Burgess, K. S.; Cameron, K. M.; Carine, M.; Chacon, J.; Clark, A.; Clarkson, J. J.; Conrad, F.; Devey, D. S.; Ford, C. S.; Hedderson, T. A. J.; Hollingsworth, M. L.; Husband, B. C.; Kelly, L. J.; Kesanakurti, P. R.; Kim, J. S.; Kim, Y.-D.; Lahaye, R.; Lee, H.-L.; Long, D. G.; Madrinan, S.; Maurin, O.; Meusnier, I.; Newmaster, S. G.; Park, C.-W.; Percy, D. M.; Petersen, G.; Richardson, J. E.; Salazar, G. A.; Savolainen, V.; Seberg, O.; Wilkinson, M. J.; Yi, D.-K.; Little, D. P.(2009). *A DNA barcodes for land plants. Proceedings of the National Academy of Science U S A, 106(31): 12794–12797.*

Ivancic, A. & V. Lebot. (2000). *The Genetics and Breeding of Taro. CIRAD, Montpellier Cedex, France.*

Kress, W.J., Wurdack, K.J., Zimmer, E.A., Weigl, A.T., & Janzen, D.H. (2005). *Use of DNA barcodes to identify flowering plants. Proceedings of the National Academy of Sciences, USA, 102, 8369–8374.*

Kress, W. J. & Erickson, D. L.(2007). *A Two-Locus Global DNA Barcode for Land Plants: The Coding rbcL Gene Complements the Non-Coding trnH–psbA spacer HpsbA Spacer Region. PLoS ONE, 2(6).*

Larkin, M. A., Blackshields, G., Brown, N. P., Chenna, R., McGettigan, P. A., McWilliam, H., Valentin, F., Wallace, I. M., Wilm, A., Lopez, R., Thompson, J. D., Gibson, T. J., & Higgins, D. G. (2007). *Clustal W and Clustal X version 2.0. Bioinformatics, 23, 2947-2948.*

Li, R., Yi, T. S., & Li, H. (2012). *Is Remusatia (Araceae) monophyletic? Evidence from three plastid regions. International Journal of Molecular Sciences, 13, 71–83.*

Mace, E. S., & Godwin, I. D. (2002). *Development and characterisation of polymorphic microsatellite markers in taro, Colocasia esculenta (L.) Schott. Genome, 45, 823–832.*

Mace, E. S., Mathur, P. N., Izquierdo, L., Hunte, D., Taylor, M. B., Singh, D., Delacy, H., Jackson, G. V. H., & Godwin, I. D. (2006). *Rationalization of taro germplasm collections in the Pacific Island region using simple sequence repeat (SSR) markers. Plant Genetic Resources, 4(3), 210–220.*

Mayo, S. J., Bogner, J. & Boyce, P. C. (1997). The genera of Araceae. Kew: Royal Botanic Gardens.

Molecular Evolutionary Genetics Analysis. Available at http://www.megasoftware.net/, accessed on March 4, 2011.

Nass, L. L., Valois, A. C. C., Melo, I. S., & Valadares-Inglis, M. C. (2001). Recursos genéticos e melhoramento- Plantas, Rondonópolis. Artigo Técnico, Fundação Mato Grosso, 1183.

Nauheimer, L., Boyce, P. C., & Renner, S. S. (2012). Giant taro and its relatives: A phylogeny of the large genus Alocasia (Araceae) sheds light on Miocene floristic exchange in the Malesian region. Molecular Phylogenetics and Evolution, 63(1), 43–51.

Newmaster, S. G., Fazekas, A. J., & Ragupathy, S. (2006). DNA barcoding in land plants: evaluation of rbcL in a multigene tiered approach. Botany, 84, 335–341.

Nunes, R. S. C., Pinhati, F. R., Golinelli, L. P., Rebouças, T. N. H., Paschoalin, V. M. F., & Silva, J. T. (2012). Polymorphic microsatellites of analysis in cultivars of taro. Horticultura Brasileira, 30(1), 106-111.

Oliveira, F. L., Guerra, J. G. M., Almeida, D. L., Ribeiro, R. L. D., Silva, E. D., Silva, V. V., & Espindola, J. A. A. (2008). Desempenho de taro em função de doses de cama de aviário, sob sistema orgânico de produção. Horticultura Brasileira, 26(2), 149-153.

Oliveira, F. L., Ribeiro, R. L. D., Silva, V. V., Guerra, J. G. M., & Almeida, D. L. (2004). Desempenho do inhame (taro) em plantio direto e no consórcio com crotalária, sob manejo orgânico. Horticultura Brasileira, 22(3), 638-641.

Pereira, F. H. F., Puiatti, M., Finger, F. L., Miranda, I.G. V., & Silva, D. J. H. (2003). Crescimento da parte aérea de dois acessos de taro e sua correlação com o rendimento de rizomas. In: 43º Congresso Brasileiro de Olericultura, Recife - PE. Brasília: Horticultura Brasileira, SOB (Sociedade de Olericultura do Brasil).

Puiatti, M., Katsumoto, R., Pereira, F. H. F., & Barrela, T. P. (2003). Crescimento de plantas e produção de rizomas de taro 'Chinês' em função do tipo de muda. Horticultura Brasileira, 21(1), 110-115.

Renner, S. S. & Zhang, L. B. (2004). Biogeography of the Pistia Clade (Araceae): based on chloroplast and mitochondrial DNA sequences and Bayesian divergence time inference. Systematic Biology, 53, 422–432.

Silva, E. E., De-Polli, H., Guerra, J. G. M., Azevedo, P. H. S., Teixeira, M. G., Espindola, J. A. A. & Almeida, M. M. T. B. (2006). Consórcio de inhame (taro) e crotalária em sistema orgânico de produção. Seropédica: Embrapa Agrobiologia. (Comunicado Técnico, 88).

Singh, D., Mace, E.S., Godwin, I.D., Mathur, P.N., Okpul, T., Taylor, M., Hunter, D., Kambuou,R., Ramanatha, R.A.O. V., Jackson, G. (2007). *Assessment and rationalization of genetic diversity of Papua New Guinea taro (Colocasia esculenta) using SSR DNA Fingerprinting. Genetic Resources Crop Evolution 55:811-822.*

Singh, S. D. R., Singh, F. F., Naresh, K. Damodaran, V., & Srivastava, R.C, (2012). *Diversity of 21 taro (Colocasia esculenta (L.) Schott) accessions of Andaman Islands. Genetic Resources Crop Evolution 59:821–829.*

Singh H. K., Parveen, I. Raghuvanshi S.Babbar, S. B.(2012). *The loci recommended as universal barcodes for plants on the basis of floristic studies may not work with congeneric species as exemplified by DNA barcoding of Dendrobium species BMC Research Notes, 5:42*

Sneath, P. H. A., & Sokal, R. R. (1973). *Unweighted Pair Group Method with Arithmetic Mean. Numerical Taxonomy. San Francisco: Freeman, 230-234.*

Sun, X.Q., Zhu, Y.-J., Guo, J.L., Peng, B., Bai, M.M., & Hang, Y.Y. (2012). *DNA Barcoding the Dioscorea in China, a Vital Group in the Evolution of Monocotyledon: Use of matK Gene for Species Discrimination. Plos One, 7(2): e32057*

TANSAO. Annual Report. (2000). *Taro Network for South East Asia and Oceania: Evaluation and breeding for rain-fed cropping systems in South East Asia and Oceania, Port Vila.*

TAROGEN. (2001). *Annual Report. AusAID/SPC Taro Genetic Resources: Conservation and Utilization. Noumea: Secretariat of the Pacific Community.*

Whitlock, B. A., Hale, A. M., & Groff, P. A. (2010). *Intraspecific Inversions Pose a Challenge for the trnH–psbA Plant DNA Barcode Plos One, 5(7): e11533.*

Xu, J., Yang, Y., Pu, Y., Ayad, W. G., & Eyzaguirre, P. B. (2001). *Genetic diversity in taro (Colocasia esculenta L. Schott, Araceae) in China: an ethnobotanical and genetic approach. Economic Botany, 55, 14-31.*

Xu, R.-Q., Tomooka, N., & Vaughan, D. A. (2000). *AFLP markers for characterizing the azuki bean complex. Crop Science, 40(3), 808-815.*

Yalu, A., Singh, D., & Yadav, S..S. (2009). *Taro Improvement and Development in Papua New Guinea - A Success Story. Asia Pacific Association of Agricultural Research Institutions (APAARI), c/FAO Regional Office for Asia and the Pacific, Bangkok, Thailand.*

Yoo, H. S, Eah, J. Y., Kim, J. S., Kim, Y. J., Min, M. S., Paek, W. K., Lee, H., & Kim, C. B. (2006). *DNA barcoding Korean birds. Molecules and Cells, 22, 323–327.*

Yu, J., Xue, J.-H., & Zhou, S.-L. (2011). *New universal matK primers for DNA barcoding angiosperms. Journal of Systematics and Evolution, 49(3), 176–181.*

Zárate, N. A. H. & Vieira, M. C. (2004). Composição nutritiva de rizomas em clones de inhame cultivados em Dourados-MS. Pesquisa Agropecuária Tropical, 34, 61-63.

Zárate, N. A. H., Vieira, M. C., Graciano, J. D., Giulani, A. R., Helmich, M., & Gomes, H. E. (2009). Produção e renda bruta de quatro clones de taro cultivados em Dourados, Estado do Mato Grosso do Sul. Acta Scientiarum, Agronomy, Maringá, 31(2), 301-305.

Zárate, N. A. H., Vieira, M. C., & Rego, N. H. (2006). Produtividade de clones de taro em função da população de plantas na época seca do pantanal sul-mato-grossense. Pesquisa Agropecuária Tropical, 36(2), 141-143.

Zárate, N.A.H.; Vieira, M. C.; Graciano, J. D.; Giuliani, A. R.; Helmich, M.; Gomes, H. E.(2009).Produção e renda bruta de quatro clones de taro cultivados em Dourados, Estado do Mato Grosso do Sul. Acta Scientiarum. Agronomy, Maringá, 31(2),301-305.

Zárate, N. A. H., Vieira, M. C., Tabaldi, L. A., Gassi, R. P., Kusano, A. M., & Maeda, A. K. M. (2012). Produção agroeconômica de taro em função do número de amontoas. Semina: Ciências Agrárias, 33(4), 1673-1680.

Zárate, N. A. H. & Yamaguti, C. Y. (1994). Curvas de crescimento de cinco clones de inhame, em solo "sempre úmido", considerando épocas de colheita, em Dourados-MS. SOB Informa, Curitiba, 13(2), 23-24.

Zhang, Z., Schwartz, S., Wagner, L., & Miller, W. (2000). A greedy algorithm for aligning DNA sequences. Journal of Computational Biology, 7, 203–214.

Zuo, L. Y. (2011). Application of DNA Barcoding in Species Identification of Colocasieae (Araceae). In: Botany. Master's thesis in Central University for Nationalities.China.

Appendix

Annex I (Sequences of the *rbcLa* locus)

```
>T37
AACGTCTGCCAATCAGCTGGTGTTAGAGATTACAAATTGACTTATTATACTCCTGAC-
TATGAGACAAAAGATACTGATATCTTGGCAGCATTCCGAGTAACTCCTCAACCCGGAG-
TTCCGCCTGAAGAAGCAGGGGCTGCAGTAGCTGCCGAATCTTCTACTGG-
TACATGGACAACTGTGTGGACTGATGGACTTACCAGTCTTGATCGTTACAAAGGACGATGC-
TACCACATCGAAGCCGTTCCTGGGGAGGAAAATCAATATATTGCTTATGTAGCTTACCCTT-
TAGACCTTTTTGAAGAAGGTTCTGTTACCAACATGTTTACTTC-
TATTGTAGGTAATGTTTTTGGGTTTAAAGCTTTACGAGCTCTACGTCTAGAGGATTT-
GCGAATTCCTCCCGCTTATTCCAAAACTTTCCAAGGCCCGCCTCACGGTATCCAAGTT-
GAAAGAGATAAATTGAACAAGTATGGTCGTCCCCTATTGGGATGTACGATTAAAC-
CAAAATTGGGATTATCCGCGAAAAACTACGGTAGAGCGGTTTATGAATGTCTCCGCGGTG-
GATTGGAATTTTACA
```

>T38
GAGCGCCATGCCAATTCAGCTGGTGTTAGAGATTACAAATTGACTTATTATACTCCTGAC-
TATGAGACAAAAGATACTGATACCTTGGCAGCATTCCGAGTAACTCCTCAACCCGGAG-
TTCCGCCTGAAGAAGCAGGGGCTGCAGTAGCTGCCGAATCTTCTACTGG-
TACATGGACAACTGTGTGGACTGATGGACTTACCAGTCTTGATCGTTACAAAGGACGATGC-
TACCACATCGAAGCCGTTCCTGGGGAGGAAAATCAATATATTGCTTATGTAGCTTACCCTT-
TAGACCTTTTTGAAGAAGGTTCTGTTACCAACATGTTTACTTC-
TATTGTAGGTAATGTTTTTGGGTTTAAAGCTTTACGAGCTCTACGTCTAGAGGATTT-
GCGAATTCCTCCCGCTTATTCCAAAACTTTCCAAGGCCCGCCTCACGGTATCCAAGTT-
GAAAGAGATAAATTGAACAAGTATGGTCGTCCCCTATTGGGATGTACGATTAAAC-
CAAAATTGGGATTATCCGCGAAAAACTACGGTAGAGCGGTTTATGAATGTCTCCGCGGTG-
GACTTGGATTTTACAA

>T39
AGGGTCATGCCATCAGTGGTGTTAGAGATTACAAATTGACTTATTATACTCCTGACTATGA-
GACAAAAGATACTGATATCTTGGCAGCATTCCGAGTAACTCCTCAACCCGGAG-
TTCCGCCTGAAGAAGCAGGGGCTGCAGTAGCTGCCGAATCTTCTACTGG-
TACATGGACAACTGTGTGGACTGATGGACTTACCAGTCTTGATCGTTACAAAGGACGATGC-
TACCACATCGAAGCCGTTCCTGGGGAGGAAAATCAATATATTGCTTATGTAGCTTACCCTT-
TAGACCTTTTTGAAGAAGGTTCTGTTACCAACATGTTTACTTC-
TATTGTAGGTAATGTTTTTGGGTTTAAAGCTTTACGAGCTCTACGTCTAGAGGATTT-
GCGAATTCCTCCCGCTTATTCCAAAACTTTCCAAGGCCCGCCTCACGGTATCCAAGTT-
GTAAGAGATAAATTGAACAAGTATGGTCGTCCCCTATTGGGATGTACGATTAAAC-
CAAAATTGGGATTATCCGCGAAAAACTACGGTAGAGCGGTTTATGAATGTCTCCGCGGTG-
GACTTGATTTTACAA

>T40
AGGGTCCAGCCAATCAGCTGGTGTTAGAGATTACAAATTGACTTATTATACTCCCTGAC-
TATGAGACAAAAGATACTGATATCCTTTGAGACAGGATTCCGAGTAACTCCTCAACCCG-
GAGTTCCGCCTGCAAGTAAGCAGGGGCTGCAGTAGCTGCCGAATCTTCTACTGG-
TACATGGACAACTGTGTGGACTGATGGACTTACCAGTCTTGATCGTTACAAAGGACGATGC-
TACCACATCGAAGCCGTTCCTGGGGAGGAAAATCAATATATTGCTTATGTAGCTTACCCTT-
TAGACCTTTTTGAAGAAGGTTCTGTTACCAACATGTTTACTTC-
TATTGTAGGTAATGTTTTTGGGTTTAAAGCTTTACGAGCTCTACGTCTAGAGGATTT-
GCGAATTCCTCCCGCTTATTCCAAAACTTTCCAAGGCCCGCCTCACGGTATCCAAGTT-
GTAAGAGATAAATTGAACAAGTATGGTCGTCCCCTATTGGGATGTACGATTAAAC-
CAAAATTGGGATTATCCGCGAAAAACTACGGTAGAGCGGTTTATGAATGTCTCCGCGGTG-
GACTTGATTTTACCAAAAACAAGAAGCGTTGGGTTGATGCTGATCAC-
TTTTTCATTCCGTCGACTCCCTCCTTCCCACCCCGACCTAACGCGGTTCTATTCGG-
TAGGGGGGGGGAATGTTT
>T41
AAGGCCTGCCATTCAGCTGGTGTTAGAGATTACAAATTGACTTATTATACTCCTGAC-
TATGAGACAAAAGATACTGATATCTTGGCAGCATTCCGAGTAACTCCTCAACCCGGAG-
TTCCGCCTGAAGAAGCAGGGGCTGCAGTAGCTGCCGAATCTTCTACTGG-
TACATGGACAACTGTGTGGACTGATGGACTTACCAGTCTTGATCGTTACAAAGGACGATGC-
TACCACATCGAAGCCGTTCCTGGGGAGGAAAATCAATATATTGCTTATGTAGCTTACCCTT-
TAGACCTTTTTGAAGAAGGTTCTGTTACCAACATGTTTACTTC-
TATTGTAGGTAATGTTTTTGGGTTTAAAGCTTTACGAGCTCTACGTCTAGAGGATTT-
GCGAATTCCTCCCGCTTATTCCAAAACTTTCCAAGGCCCGCCTCACGGTATCCAAGTT-
GAAAGAGATAAATTGAACAAGTATGGTCGTCCCCTATTGGGATGTACGATTAAAC-
CAAAATTGGGATTATCCGCGAAAAACTACGGTAGAGCGGTTTATGAATGTCTCCGCGGTG-
GACTGGATTTTACA

>T42
GAAGGGCCTGGCAGTCAGTGGTGTTAGAGATTACAAATTGACTTATTATACTCCTGAC-
TATGAGACAAAAGATACTGATACCGTAGGAGGATTCCGAGTAACTCCTCAACCCGGAG-
TTCCGCCTGCCAATGAAGCAGGGGCTGCAGTAGCTGCCGAATCTTCTACTGG-
TACATGGACAACTGTGTGGACTGATGGACTTACCAGTCTTGATCGTTACAAAGGACGATGC-

```
TACCACATCGAAGCCGTTCCTGGGGAGGAAAATCAATATATTGCTTATGTAGCTTACCCTT-
TAGACCTTTTTGAAGAAGGTTCTGTTACCAACATGTTTACTTC-
TATTGTAGGTAATGTTTTTGGGTTTAAAGCTTTACGAGCTCTACGTCTAGAGGATTT-
GCGAATTCCTCCCGCTTATTCCAAAACTTTCCAAGGCCCGCCTCACGGTATCCAAGTT-
GAAAGAGATAAATTGAACAAGTATGGTCGTCCCCTATTGGGATGTACGATTAAAC-
CAAAATTGGGATTATCCGCGAAAAACTACGGTAGAGCGGTTTATGAATGTCTCCGCGGTG-
GACTTGGATTTTACAACCAACATTCAGTGGGGGTGACGCTGGTGCCGAG-
TACATTCCTGAGTCTACCGGTGTCTTCACCACTACTTAAAGGCCGGTCTAATCTGAAGGGGGGAGGAAA
```

>T43
```
GGGGCTTTAACCGTCATCTACGTAGTTTTTCGCGGATAATCCCAATTTTGGTTCTAATCG-
TACATCCCAATAGGGGACGACCATCCATATAGCAATTTATCTCTTTCAACTTGGATAC-
CGTGAGGCGGGCCTTGGAAAGTTTTGGAATAAGCGGGAGGAATTCGCAAATCCTCTAGAC-
GTAGAGCTCGTAAAGCTTTAAACCCAAAAACATTACCTACAATAGAAGTAAACATGTTGG-
TAACAGAACCTTCTTCAAAAAGGTCTAAAGGGTAAGCTACATAAGCAATATATTGAT-
TTTCCTCCCCAGGAACGGCTTCGATGTGGTAGCATCGTCCTTTGTAACGATCAAGACTGG-
TAAGTCCATCAGTCCACACAGTTGTCCATGTACCAGTAGAAGATTCGGCAGCTACTG-
CAGCCCCTGCTTCTTCAGGCGGAACTCCGGGTTGAGGAGTTACTCGGAATGCTGCCAAGA-
TATCAGTATCTTTTGTCTCATAGTCAGGAGTATAATAAGTCAATTTGTAATCTCTAACAC-
CAGCTTTGAATCCAGCACTTGCTTTAGTCTCTGTTTGGGGTGACATAGCA
```

Annex II (Sequences of the *matK* locus)

>T37
```
GCGTCTTTGGGTAGTTCCATTTAATTGTGTATCAGATATACTAATACCCTATCCCG-
TACATCTAGAAATCTTGGTTCAAATTCTAAAATGCTGGATACAA-
GATGTTCCTTCTTTACATTTATTATGATTCTTTTTTCACGAATATTATAATTGGAA-
TAATCTCATTACTCCAAAGAAATCTAACTATTATGGGTTTTCGAAAGAGAATCCAA-
GACTTTTTTTGTTCCTATATAATTCTTATGTAGTTGAATGCGAATCCATATTAG-
TTTTTCTCCGTAAACAATCCTCTTATTTACAATCAACATCTTCTGGAACCTTTCTT-
GAGCGAACACATTTCTATGAAAAAATAGAACAACATCTCGTAGTACTTTGTTGTAATGAT-
TTGCAGAAAACCCTATGGTTGTTCAAGGATCCTTTCATACATTATGTTAGA-
TATCAAGGAAAATCAATTCTGGCTTCAAAAGGGACTCATCTTCTGATGAA-
GAAATGGAAATCTTACTTTGTCAATTTTTGGCAATGTCATTTTCACTTTT-
GGTCTCAACCCAGTAGGATCCACATAAGCCAATTCTCCAATTTTTCTTTC-
TATTTTCTGGGTTATCTTTCAAGTGTACCAATAAATCCTTCAGCGGTAAAGAGTCAAATGC-
TAGAGAGTTCTTTTTTAATAGATACTGTTACTAAAAAATTCGAAAC-
TATAGTTCCAATTATTCCAATGATTGGATCATTGTCAAAAGCTAAATTTTGTAAC-
GTATCGGGGAATCCTATTAGTAAGCCAGTTTGGACCGATTTGTCAGATTCTGA-
TATTATTGATCGATTTGGTCGGATATGTAGAAATCTTTCTCATTATTACAG-
TGGGTCTTCAAAAAAACAAAGTTTGTATCGAATAAAGTATATACTTCGACTCGTGTGTGAAAAAA
```

>T38
```
GAGTTGATTCGTTAACTTTGTTTTTTTGAAGACCCACTGTAATAATGAGAAAGATTTC-
TACATATCCGACCAAATCGATCAATAA-
TATCAGAATCTGACAAATCGGTCCAAACTGGCTTACTAATAGGATTCCCCGATAC-
GTTACAAAATTTAGCTTTTGACAATGATCCAATCATTGGAATAATTGGAAC-
TATAGTTTCGAATTTTTTAGTAACAGTATCTATTAAAAAAGAACTCTCTAGCATTT-
GACTCTTTACCGCTGAAGGATTTATTGGTACACTTGAAAGATAACCCAGAAAATAGAAA-
GAAAAATTGGAGAATTGGCTTATGTGGATCCTACTGGGTTGAGAC-
CAAAAGTGAAAATGACATTGCCAAAAATTGACAAAGTAAGATTTCCATTTCTTCATCAGAA-
GATGAGTCCCTTTTGAAGCCAGAATTGATTTTCCTTGA-
TATCTAACATAATGTATGAAAGGATCCTTGAACAACCATAGGGTTTTCTG-
CAAATCATTACAACAAAGTACTACGAGATGTTGTTCTATTTTTTCATA-
GAAATGTGTTCGCTCAAGAAAGGTTCCAGAAGATGTTGATTGTAAATAAGAG-
GATTGTTTACGGAGAAAAACTAATATGGATTCGCATTCAACTACATAAGAATTATA-
TAGGAACAAAAAAAGTCTTGGATTCTCTTTCGAAAACCCATAATAGTTAGATTTCTTT-
```

GGAGTAATGAGATTATTCCAATTATAATATTCGTGAAAAAAGAATCGTAATAAATGTAAA-
GAAGGAACATCTTGTATCCAGCATTGTAGAATTTGAACCAAGATTTCTAGATGTACGGGA-
TAGGGTATTAGTATATCTGATACACAATTTAAATGTGATAATTTGTCCTCAAAATGAG-
TGAATAAAAATTGGGGAGG

>T39
ACCTTTTGGGGCAATTCCATTTAAATTGTGTATCAGATATACTAATACCCTATCCCG-
TACATCTAGAAATCTTGGTTCCAATTCTACAATGCTGGATACAA-
GATGTTCCTTCTTTACATTTATTACGATTCTTTTTTCACGAATATTATAATTGGAA-
TAATCTCATTACTCCAAAGAAATCTAACTATTATGGGTTTTCGAAAGAGAATCCAA-
GACTTTTTTTGTTCCTATATAATTCTTATGTAGTTGAATGCGAATCCATATTAG-
TTTTTCTCCGTAAACAATCCTCTTATTTACAATCAACATCTTCTGGAACCTTTCTT-
GAGCGAACACATTTCTATGAAAAAATAGAACAACATCTCGTAGTACTTTGTTGTAATGAT-
TTGCAGAAAACCCTATGGTTGTTCAAGGATCCTTTCATACATTATGTTAGA-
TATCAAGGAAAATCAATTCTGGCTTCAAAAGGGACTCATCTTCTGATGAA-
GAAATGGAAATCTTACTTTGTCAATTTTTGGCAATGTCATTTTCACTTTT-
GGTCTCAACCCAGTAGGATCCACATAAGCCAATTCTCCAATTTTTCTTTC-
TATTTTCTGGGTTATCTTTCAAGTGTACCAATAAATCCTTCAGCGGTAAAGAGTCAAATGC-
TAGAGAGTTCTTTTTTAATAGATACTGTTACTAAAAAATTCGAAAC-
TATAGTTCCAATTATTCCAATGATTGGATCATTGTCAAAAGCTAAATTTTGTAAC-
GTATCGGGGAATCCTATTAGTAAGCCAGTTTGGACCGATTTGTCAGATTCTGA-
TATTATTGATCGATTTGGTCGGATATGTAGAAATCTTTCTCATTATTACAG-
TGGGTCTTCAAAAAAACAAAGTTTGTATCGAATAAAGTATATACTTCGACTTCGTGGCTGAAGAAAAGA

>T40
CACATGTATTTGATCCAGTTGTTTTTTCTGAAGACCCACTGTAATTAATGAGAAAGATTTC-
TACATATCCGACCAAATCGATCAATTA-
TATCAGAATCTGACAAATCGGTCCAAACTGGCTTACTAATAGGATTCCCCGATAC-
GTTACAAAATTTAGCTTTTGACAATGATCCAATCATTGGAATAATTGGAAC-
TATAGTTTCGAATTTTTTAGTAACAGTATCTATTAAAAAAGAACTCTCTAGCATTT-
GACTCTTTACCGCTGAAGGATTTATTGGTACACTTGAAAGATAACCCAGAAAATAGAAA-
GAAAAATTGGAGAATTGGCTTATGTGGATCCTACTGGGTTGAGAC-
CAAAAGTGAAAATGACATTGCCAAAAATTGACAAAGTAAGATTTCCATTTCTTCATCAGAA-
GATGAGTCCCTTTTGAAGCCAGAATTGATTTTCCTTGA-
TATCTAACATAATGTATGAAAGGATCCTTGAACAACCATAGGGTTTTCTG-
CAAATCATTACAACAAAGTACTACGAGATGTTGTTCTATTTTTTCATA-
GAAATGTGTTCGCTCAAGAAAGGTTCCAGAAGATGTTGATTGTAAATAAGAG-
GATTGTTTACGGAGAAAAACTAATATGGATTCGCATTCAACTACATAAGAATTATA-
TAGGAACAAAAAAAGTCTTGGATTCTCTTTCGAAAACCCATAATAGTTAGATTTCTTT-
GGAGTAATGAGATTATTCCAATTATAATATTCGTGAAAAAAGAATCGTAATAAATGTAAA-
GAAGGAACATCTTGTATCCAGCATTGTAGAATTTGAACCAAGATTTCTAGATGTACGGGA-
TAGGGTATTAGTATATCTGATACACAATTTAAATGTGATAATTT-
GTCCTCAAAATGAGTAAAAAAAAAAAGAAGAGGGCGC

>T41
ACCGTTTGGAGCCAAGTTCTATTTAATTGTGTATCAGATATACTAATACCCTATCCCG-
TACATCTAGAAATCTTGGTTCCCATCATAAGAATGCTGGATACAA-
GATGTTCCTTCTTTACATTTATCACTGATTCTTTTTTCACGAATATTATAATTGGAA-
TAATCTCATTACTCCAAAGAAATCTAACTATTATGGGTTTTCGAAAGAGAATCCAA-
GACTTTTTTTGTTCCTATATAATTCTTATGTAGTTGAATGCGAATCCATATTAG-
TTTTTCTCCGTAAACAATCCTCTTATTTACAATCAACATCTTCTGGAACCTTTCTT-
GAGCGAACACATTTCTATGAAAAAATAGAACAACATCTCGTAGTACTTTGTTGTAATGAT-
TTGCAGAAAACCCTATGGTTGTTCAAGGATCCTTTCATACATTATGTTAGA-
TATCAAGGAAAATCAATTCTGGCTTCAAAAGGGACTCATCTTCTGATGAA-
GAAATGGAAATCTTACTTTGTCAATTTTTGGCAATGTCATTTTCACTTTT-
GGTCTCAACCCAGTAGGATCCACATAAGCCAATTCTCCAATTTTTCTTTC-
TATTTTCTGGGTTATCTTTCAAGTGTACCAATAAATCCTTCAGCGGTAAAGAGTCAAATGC-
TAGAGAGTTCTTTTTTAATAGATACTGTTACTAAAAAATTCGAAAC-

```
TATAGTTCCAATTATTCCAATGATTGGATCATTGTCAAAAGCTAAATTTTGTAAC-
GTATCGGGGAATCCTATTAGTAAGCCAGTTTGGACCGATTTGTCAGATTCTGA-
TATTATTGATCGATTTGGTCGGATATGTAGAAATCTTTCTCATTATTACAG-
TGGGTCTTCAAAAAAACAAAGTTTGTATCGAATAAAGTATATACTCATGTTGCTGGAGAGA-
GAGAAAAAAAAA
```

>T42
```
GACTAGACTTTGAACCACCTGTTTTTTCTGAAGACCACTGGTAATAATGAAGAAGAATTTC-
TACATATCCGACCAAATCGATCAACAT-
ATCAGAATCTGACAAATCGGTCCAAACTGGCTTACTAATAGGATTCCCCGATAC-
GTTACAAAATTTAGCTTTTGACAATGATCCAATCATTGGAATAATTGGAAC-
TATAGTTTCGAATTTTTTAGTAACAGTATCTATTAAAAAAGAACTCTCTAGCATTT-
GACTCTTTACCGCTGAAGGATTTATTGGTACACTTGAAAGATAACCCAGAAAATAGAAA-
GAAAAATTGGAGAATTGGCTTATGTGGATCCTACTGGGTTGAGAC-
CAAAAGTGAAAATGACATTGCCAAAAATTGACAAAGTAAGATTTCCATTTCTTCATCAGAA-
GATGAGTCCCTTTTGAAGCCAGAATTGATTTTCCTTGA-
TATCTAACATAATGTATGAAAGGATCCTTGAACAACCATAGGGTTTTCTG-
CAAATCATTACAACAAAGTACTACGAGATGTTGTTCTATTTTTTCATA-
GAAATGTGTTCGCTCAAGAAAGGTTCCAGAAGATGTTGATTGTAAATAAGAG-
GATTGTTTACGGAGAAAAACTAATATGGATTCGCATTCAACTACATAAGAATTATA-
TAGGAACAAAAAAAGTCTTGGATTCTCTTTCGAAAACCCATAATAGTTAGATTTCTTT-
GGAGTAATGAGATTATTCCAATTATAATATTCGTGAAAAAAGAATCGTAATAAATGTAAA-
GAAGGAACATCTTGTATCCAGCATTGTAGAATTTGAACCAAGATTTCTAGATGTACGGGA-
TAGGGTATTAGTATATCTGATACACAATTTAAATGTGATAATTTGTCCTCAAATGAG-
TTTAATATTAGGGGGGGGGG
```

>T43
```
AAAGTTTGGGGAAAAATTCCTTTAATTGTGTATCAGATATACTAATACCCTATCCCG-
TACATCTAGAAATCTTGGTCCTTAAACTATATGCTGGATACAA-
GATGTTCCTTCTTTACATTTACTTTCTAATCTTTTTTCACGAATATTATAATTGGAA-
TAATCTCATTACTCCAAAGAAATCTAACTATTATGGGTTTTCGAAATTAGAATCCAA-
GACTTTTTTTGTTCCTATATAATTCTTATGTAGTTGAATGCGAATCCATATTAG-
TTTTTCTCCGTAAACAATCCTCTTATTTACAATCAACATCTTCTGGAACCTTTCTT-
GAGCGAACACATTTCTATGAAAAAATAGAACAACATCTCGTAGTACTTTGTTGTAATGAT-
TTGCAGAAAACCCTATGGTTGTTCAAGGATCCTTTCATACATTATGTTAGA-
TATCAAGGAAATCAATTCTGGCTTCAAAAGGGACTCATCTTCTGATGAA-
GAAATGGAAATCTTACTTTGTCAATTTTTGGCAATGTCATTTTCACTTTT-
GGTCTCAACCCAGTAGGATCCACATAAGCCAAGTCTCCAATTATTCTTTC-
TATTTTCTGGGTTATCTTTCAAGTGTACCAATAAATCCTTCAGCGGTAAAGAGTCAAATGC-
TAGAGAGTTCATTTTTAATAGATACTGTTACTAAAAAATTCGACAC-
TATAGTTCCAATTATTTCAATGATTGGATCATTGTCAAAAGCTAAATTTTGTAAC-
GTATCGCGGAATCCATATTAGTAAGCCAGTTTGGACCGACTTGTCAGATTCTGA-
TATTATTGATCGATTTGGTCGGGATATGTAGAAATCTTTTCTCATTATTACAGTT-
GGGTCTTCAAAATAACAAAGATAGTATCGAATAAAGTATATACTTCGACTTTTCGTGTGCAAAAA
```

Annex III (Sequences of the *pbsA-trnH* intergenic spacer)

>T37
```
ATATAATATTATATATATATAGAATATTATAATATTTCTATATTTATTATATAATATTTCTATTATTATTTC
TTTTTTATATTAATATTCTATTTTATTTATTTATATTTATTTATTATTTCTATTTTATTTTTTATATATTTT
ATTTTTCTTTTTGTTTTATATTTAACTTTTTTTATTCTTATTCTTTGCTTTGTATTTTTTTGTATTTAGTTT
TTCGTTCATACATTTATTTTGTATTTCAAATAAATAAAAGATTTGATTTCAAAGCCAAAGAAGTACATAATT
ACGGATTGGTAAAAGCCAAAGTATGCTAAAATATTAAAATAAAAGTACAAATGTCAAAAATGCTTATGTGGA
CAAAATCCCAGTGAATCAAAAAAAGGAGTAATACCCAAACCTCCTCATTAGAGGTGTGGTATTATTCCTTCAA
CAATTCCTATACACTAAGACAAAATGTCTTATCCATTTGTAGATGGAACTTCAACAGCAGCTAAGTCTAGAG
GGAAGTTGTGAGCATTACGTTCATGCATTACTTCCATCCGAGTGTT
```

>T38
ATATATATATATATATATATAATATTATATATATATAGAATATTATAATATTTCTATATTTATTATATAATA
TTTCTATTATTATTTCTTTTTTATATTAATATTCTATTTTATTTATTTATATTTATTTATTATTTCTATTTT
ATTTTTTATATATTTTATTTTTCTTTTTGTTTTATATTTAACTTTTTTATTCTTATTCTTTGCTTTGTATTT
TTTTTGTATTTAGTTTTTCGTTCATACATTTATTTTGTATTTCAAATAAATAAAAGATTTGATTTCAAAGCC
AAAGAAGTACATAATTACGGATTGGTAAAAGCCAAAGTATGCTAAAATATGAAAATAAAAGTACAAATGTCA
AAAATGCTTATGTGGACAAAATCCCAGTGAATCAAAAAAAGGAGTAATACCCATAGTAGGGCATTAGAGGTT
TGGTATTATTCCTTCAACAATTCCTAGACTCAAAGACAAAATGTCTTATCCATTTGTAGATGGAACTTCAAC
AGCAGCTAAGTCTAGAGGAACGAGAACTTAAATCGGA

>T39
ATTCTAAGAATATATATATAATATTATATATATATAGAATATTATAATATTTCTATATTTATTATATAATAT
TTCTATTATTATTTCTTTTTTATATTAATATTCTATTTTATTTATTTATATTTATTTATTATTTCTATTTTA
TTTTTTATATATTTTATTTTTCTTTTTGTTTTATATTTAACTTTTTTATTCTTATTCTTTGCTTTGTATTTT
TTTTGTATTTAGTTTTTCGTTCATACATTTATTTTGTATTTCAAATAAATAAAAGATTTGATTTCAAAGCCA
AAGAAGTACATAATTACGGATTGGTAAAATCCTAAGTATGCTAAAATATTAAAATAAAAGTACAAATGTCAA
AAATGCTTATGTGGACAAAATCCCAGTGAATCAAAAAAAGGAGTAATACCACACCTCCTCATTAGAGGTTTG
GTATTATTCCTTCAACAATTCCTATACACTAAGACAAAATGTCTTATCCATTTGTAGATGGAACTTCAACAG
CAGCTAAGTCTAGAGGGAAGTTGTGAGCATTACGTTCATGCATTACTTCCATCCGGGGGGA

>T40
ATATATATATATATATATATAATATTATATATATATAGAATATTATAATATTTCTATATTTATTCTATAATA
TTTCTATTATTATTTCTTTTTTATATTAATATTCTATTTTATTTATTTATATTTATTTATTATTTCTATTTT
ATTTTTTATATATTTTATTTTTCTTTTTGTTTTATATTTAACTTTTTTATTCTTATTCTTTGCTTTGTATTT
TTTTTGTATTTAGTTTTTCGTTCATACATTTATTTTGTATTTCAAATAAATAAAAGATTTGATTTCAAAGCC
AAAGAAGTACATAATTACGGATTGGTAAAAGCCAAAGTATGCTAAAAATAAAAATAAAAGTACAAATGTCAA
AAACGCTTATGTGGACAAAATTCCCCACTTCTATACATCCGCGTGATCATCTCGTAGGCTCCGCTTGATCCA
CTTGGCTACATCCGCTTGATCCACTTACTAGATCCGCTTGATCCACTTGGCTACATCCGCCTGATCCACTGG
CTACACCGCCTGTACATGATATTTTCCTCGGC

>T41
ACTTGGCTACATCCGCCAGAACCACTTGGCTACTTCCGCCTGATCCACTTGGATACATCCGCCTGATTCTTT
TTTATATAAATATTCTATTTTATTTATTTATATTTATTTATTATTTCTATTTTATTTTTTATATATTTTATT
TTTCTTTTTGTTTTATATTTAACTTTTTTATTCTTATTCTTTGCTTTGTATTTTTTTTGTATTTAGTTTTTC
GTTCATACATTTATTTTGTATTTCAAATAAATAAAAGATTTGATTTCAAAGCCAAAGAAGTACATAATTACG
GATTGGTAAAAGCCAAAGTATGCTAAAATATTAAAATAAAAGTACAAATGTCAAAAATGCTTATGTGGACAA
AATCCCAGTGAATCAAAAAAAGGAGTAATACCAAACCTCCTCATTAGAGGTTTGGTATTATTCCTTCAACAA
TTCCTATACACTAAGACAAAATGTCTTATCCATTTGTAGATGGAACTTCAACAGCAGCTAAGTCTAGAGGGA
AGTTGTGAGCATTACGTTCATGCATTACTTCCTCCAAGGGGTTAGACACACCCCCCCCCCAGAGGA
>T42
ATATATATATATATATATATAATATTATATATATATAGAATATTATAATATTTCTATATTTATTATATAATA
TTTCTATTATTATTTCTTTTTTATATTAATATTCTATTTTATTTATTTATATTTATTTATTATTTCTATTTT
ATTTTTTATATATTTTATTTTTCTTTTTGTTTTATATTTAACTTTTTTATTCTTATTCTTTGCTTTGTATTT
TTTTTGTATTTAGTTTTTCGTTCATACATTTATTTTGTATTTCAAATAAATAAAAGATTTGATTTCAAAGCC
AAAGAAGTACATAATTACGGATTGGTAAAAGCCAAAGTATGCTAAAATATTAAAATTAAAGTTCAAATGTCA
AAAATAGTTATTTGAACACAAGGTTAGCACTTCCATACCAAGGTTAGCACTTCCAAACCAAGGTTAGCACTT
CCATACCAAGGTTAGCACTTCCATACCATTAAGGGCACTTCCATACCAAGGTTAGCACTTCCATACCAAGGT
TAGCACTTCCATCCAAGTTCTAAAACCTTCCTGGGT

>T43
ATATATATATATATATATATATAGAATATTATATATATATAGAATATTATAATATTTCTATATTTATTAT
ATAATATTTCTATTATTATTTCTTTTTTATATTAATATTCTATTTTATTTATTTATATTTATTTATTATTTC
TATTTTATTTTTTATATATTTTATTTTTCTTTTTGTTTTATATTTAACTTTTTTATTCTTATTCTTTGCTTT
GTATTTTTTTTGTATTTAGTTTTTCGTTCATACATTTATTTTGTATTTCAAATAAATAAAAGATTTGATTTC
AAAGCCAAAGAAGTACATAATTACGGATTGGTAAAAGCCAAAGTATGCTAAAATATTAAAATAAAAGTACAA
ATGTCAAAAATGCTTATGTGGACAAAATCCCAGTGAATCAAAAAAAGGAGTAATACCAAACCTCCTCATTAG
AGGTTTGGTATTATTCCTTCAACAATTCCTATACGAGAAGACAAAATGTCTTATCCATTTGTAGATGGAACT
TCAACAGCAGCTAAGTCTAGATGTAAGCTGATCCTTGTACCG

Annex IV (Alignment of the *rbcL* locus sequences)

```
CLUSTAL 2.1 multiple sequence alignment
T37    -AACGTC-TGCCAA-TCAGCTGGTGTTAGAGATTACAAATTGACTTATTATACTCC-TGA
T41    -AAGGCC-TGCCAT-TCAGCTGGTGTTAGAGATTACAAATTGACTTATTATACTCC-TGA
T38    GAGCGCCATGCCAATTCAGCTGGTGTTAGAGATTACAAATTGACTTATTATACTCC-TGA
T39    -AGGGTCATGCCA--TCAG-TGGTGTTAGAGATTACAAATTGACTTATTATACTCC-TGA
T40    -AGGGTCCAGCCAA-TCAGCTGGTGTTAGAGATTACAAATTGACTTATTATACTCCCTGA
T42    GAAGGGCCTGGCAG-TCAG-TGGTGTTAGAGATTACAAATTGACTTATTATACTCC-TGA
T43    -GGGGCTTTAACCG-TCATCTA-CGTAGTTTTTCGCGGATAATCCCAATTTTGGTTCTAA
                 *      * *** *    **        *  *  **   *  *  * *      * *


T37    CTATGAGAC--AAAAGATACTGATATCTT----GGCAGC--ATTCCGAGTAACTCCTCAA
T41    CTATGAGAC--AAAAGATACTGATATCTT----GGCAGC--ATTCCGAGTAACTCCTCAA
T38    CTATGAGAC--AAAAGATACTGATACCTT----GGCAGC--ATTCCGAGTAACTCCTCAA
T39    CTATGAGAC--AAAAGATACTGATATCTT----GGCAGC--ATTCCGAGTAACTCCTCAA
T40    CTATGAGAC--AAAAGATACTGATATCCTTTGAGACAGG--ATTCCGAGTAACTCCTCAA
T42    CTATGAGAC--AAAAGATACTGATACCGT----AGGAGG--ATTCCGAGTAACTCCTCAA
T43    TCGTACATCCCAATAGGGGACGACCATCCATATAGCAATTTATCTCTTTCAACTTGGATA
             *   *   ** **    **          *    ** *    ****       *


T37    CCCGGAGTTCCGCCTGA-AG-----AAGCAGGGGCTGCAGTAGCTGCCGAATCTTCTACT
T41    CCCGGAGTTCCGCCTGA-AG-----AAGCAGGGGCTGCAGTAGCTGCCGAATCTTCTACT
T38    CCCGGAGTTCCGCCTGA-AG-----AAGCAGGGGCTGCAGTAGCTGCCGAATCTTCTACT
T39    CCCGGAGTTCCGCCTGA-AG-----AAGCAGGGGCTGCAGTAGCTGCCGAATCTTCTACT
T40    CCCGGAGTTCCGCCTGC-AAGT---AAGCAGGGGCTGCAGTAGCTGCCGAATCTTCTACT
T42    CCCGGAGTTCCGCCTGCCAATG---AAGCAGGGGCTGCAGTAGCTGCCGAATCTTCTACT
T43    CCGTGAGGCGGGCCTTGGAAAGTTTTGGAATAAGCGGGAGGAATTCGCAAATCCTCTAG-
       **  ***   ****    *       * *   ** * **  *   *  * **** ****


T37    GGTACATGGACAACTGTGTGGACTGATGGACTTACCAGTCTTGATCGTTACAAAGGACGA
T41    GGTACATGGACAACTGTGTGGACTGATGGACTTACCAGTCTTGATCGTTACAAAGGACGA
T38    GGTACATGGACAACTGTGTGGACTGATGGACTTACCAGTCTTGATCGTTACAAAGGACGA
T39    GGTACATGGACAACTGTGTGGACTGATGGACTTACCAGTCTTGATCGTTACAAAGGACGA
T40    GGTACATGGACAACTGTGTGGACTGATGGACTTACCAGTCTTGATCGTTACAAAGGACGA
T42    GGTACATGGACAACTGTGTGGACTGATGGACTTACCAGTCTTGATCGTTACAAAGGACGA
T43    -----ACGTAGAGCT-CGTAAAGCTTTAAACCCAAAAACATTACCTACAATAGAAGTAAA
            *  * * **   **  *    *  **  *  *  **       *  *  *     *


T37    TGCTACCACATCGAAGCCGTTCCTGGGGAGGAA-AATCAATATATTGCTTATGTAGCTTA
T41    TGCTACCACATCGAAGCCGTTCCTGGGGAGGAA-AATCAATATATTGCTTATGTAGCTTA
T38    TGCTACCACATCGAAGCCGTTCCTGGGGAGGAA-AATCAATATATTGCTTATGTAGCTTA
T39    TGCTACCACATCGAAGCCGTTCCTGGGGAGGAA-AATCAATATATTGCTTATGTAGCTTA
T40    TGCTACCACATCGAAGCCGTTCCTGGGGAGGAA-AATCAATATATTGCTTATGTAGCTTA
T42    TGCTACCACATCGAAGCCGTTCCTGGGGAGGAA-AATCAATATATTGCTTATGTAGCTTA
T43    CATGTTGGTAACAGAACCTTCTTCAAAAAGGTCTAAAGGGTAAGCTACATAAGCAATATA
          * *  * **  *      ***   **   **   * * ** * *     **


T37    CCCTTTAGACCTTTTTGAAGAAGGTTCTGTTACCAACATGTTTACTTCTATTGTAGGTAA
T41    CCCTTTAGACCTTTTTGAAGAAGGTTCTGTTACCAACATGTTTACTTCTATTGTAGGTAA
T38    CCCTTTAGACCTTTTTGAAGAAGGTTCTGTTACCAACATGTTTACTTCTATTGTAGGTAA
T39    CCCTTTAGACCTTTTTGAAGAAGGTTCTGTTACCAACATGTTTACTTCTATTGTAGGTAA
T40    CCCTTTAGACCTTTTTGAAGAAGGTTCTGTTACCAACATGTTTACTTCTATTGTAGGTAA
T42    CCCTTTAGACCTTTTTGAAGAAGGTTCTGTTACCAACATGTTTACTTCTATTGTAGGTAA
T43    TTGATTTT-CCTCCCCAGGAACGGCTTCGATGTGGTAGCATCGTCCTTTGTAACGATCAA
         **   ***     *  ** *   * *         *   * ** *       **
```

```
T37    TGTTTTTGGGTTTAAAGCTTTACG-AGCTCTACGTCTAGAGGATTTGCGAATTCCTC--C
T41    TGTTTTTGGGTTTAAAGCTTTACG-AGCTCTACGTCTAGAGGATTTGCGAATTCCTC--C
T38    TGTTTTTGGGTTTAAAGCTTTACG-AGCTCTACGTCTAGAGGATTTGCGAATTCCTC--C
T39    TGTTTTTGGGTTTAAAGCTTTACG-AGCTCTACGTCTAGAGGATTTGCGAATTCCTC--C
T40    TGTTTTTGGGTTTAAAGCTTTACG-AGCTCTACGTCTAGAGGATTTGCGAATTCCTC--C
T42    TGTTTTTGGGTTTAAAGCTTTACG-AGCTCTACGTCTAGAGGATTTGCGAATTCCTC--C
T43    GACTGGTAAGTCCATCAGTCCACACAGTTGTCCATGTACCAG--TAGAAGATTCGGCAGC
         *    *   **  *       *   **   **  *   *   *  **     *    *  *    ****    *   *

T37    CGCTTATTCCAAAACTTTCCAAGGCCCGCCTCACGGTATCCAAGTTGAAAGAGATAAATT
T41    CGCTTATTCCAAAACTTTCCAAGGCCCGCCTCACGGTATCCAAGTTGAAAGAGATAAATT
T38    CGCTTATTCCAAAACTTTCCAAGGCCCGCCTCACGGTATCCAAGTTGAAAGAGATAAATT
T39    CGCTTATTCCAAAACTTTCCAAGGCCCGCCTCACGGTATCCAAGTTGTAAGAGATAAATT
T40    CGCTTATTCCAAAACTTTCCAAGGCCCGCCTCACGGTATCCAAGTTGTAAGAGATAAATT
T42    CGCTTATTCCAAAACTTTCCAAGGCCCGCCTCACGGTATCCAAGTTGAAAGAGATAAATT
T43    TACTGCAGCCCCTGCTTCTTCAGGCGGAACTCCGGGTTGAGGAGTTACTCGGAATGCTGC
         **      **      ***       ****      ***   ***      ****     *    **

T37    GAACA----AGTATGGTC-GTCCCCTATTGGGATGTACGATTAAACCAAAATTGGGATTA
T41    GAACA----AGTATGGTC-GTCCCCTATTGGGATGTACGATTAAACCAAAATTGGGATTA
T38    GAACA----AGTATGGTC-GTCCCCTATTGGGATGTACGATTAAACCAAAATTGGGATTA
T39    GAACA----AGTATGGTC-GTCCCCTATTGGGATGTACGATTAAACCAAAATTGGGATTA
T40    GAACA----AGTATGGTC-GTCCCCTATTGGGATGTACGATTAAACCAAAATTGGGATTA
T42    GAACA----AGTATGGTC-GTCCCCTATTGGGATGTACGATTAAACCAAAATTGGGATTA
T43    CAAGATATCAGTATCTTTTGTCTCATAGTCAGGAGTATAAT--AAGTCAATTTGTAATCT
         ** *       *****   *   *** ** **   *    ***    **   **   ** *** **

T37    TCCGCGAAAAACTACGGTAGAGCGGTTTATGAATGTCTCCGCGGTGGATTGGAATTTTAC
T41    TCCGCGAAAAACTACGGTAGAGCGGTTTATGAATGTCTCCGCGGTGGACTGGA-TTTTAC
T38    TCCGCGAAAAACTACGGTAGAGCGGTTTATGAATGTCTCCGCGGTGGACTTGGATTTTAC
T39    TCCGCGAAAAACTACGGTAGAGCGGTTTATGAATGTCTCCGCGGTGGACTTG-ATTTTAC
T40    TCCGCGAAAAACTACGGTAGAGCGGTTTATGAATGTCTCCGCGGTGGACTTG-ATTTTAC
T42    TCCGCGAAAAACTACGGTAGAGCGGTTTATGAATGTCTCCGCGGTGGACTTGGATTTTAC
T43    CTAACACCAGCTTTGAATCCAGCACTTGCTTTA-GTCTCTGTT-TGGGGTGACATAGCA-
         *    *    *    *   ***   **   *   * *****  *    ***   *     *    *
```

Annex V (Alignment of the *matK* locus sequences)

```
CLUSTAL 2.1 multiple sequence alignment
T39    ACCTTTTGGGGC--AATTCCATTTAAATTGTGTATCAGATATACTA-ATACCCTATCCCG
T41    ACCGTTTGGAGCCAAGTTCTATTTAA-TTGTGTATCAGATATACTA-ATACCCTATCCCG
T37    GCGTCTTTGGGT--AGTTCCATTTAA-TTGTGTATCAGATATACTA-ATACCCTATCCCG
T43    AAAGTTTGGGGAAAAATTCC-TTTAA-TTGTGTATCAGATATACTA-ATACCCTATCCCG
T38    CTTTGTTTTTTTGAAGACCCACTGTAAT-AATGAGAAAGATTTCTACATATCCGACCAAA
T40    GTTGTTTTTTCTGAAGACCCACTGTAATTAATGAGAAAGATTTCTACATATCCGACCAAA
T42    CCTGTTTTTTCTGAAGACCACTGGTAATAATGAGAAGAATTTCTACATATCCGACCAAA
         **        *    *         * *       *   *    * *** *** ** * *

T39    TACATCTAG-AAATCTTGGTTC--CAATTCTAC-AATGCTGGATA--CAAGATGTTCCTT
T41    TACATCTAG-AAATCTTGGTTC--CCATCATAAGAATGCTGGATA--CAAGATGTTCCTT
T37    TACATCTAG-AAATCTTGGTTC--AAATTCTAA-AATGCTGGATA--CAAGATGTTCCTT
T43    TACATCTAG-AAATCTTGGTCC--TTAAACTAT--ATGCTGGATA--CAAGATGTTCCTT
T38    TCGATCAATAATATCAGAATCTGACAAATCGGTCCAAACTGGCTTACTAATAGGATTCCC
T40    TCGATCAATTATATCAGAATCTGACAAATCGGTCCAAACTGGCTTACTAATAGGATTCCC
T42    TCGATCAAC-ATATCAGAATCTGACAAATCGGTCCAAACTGGCTTACTAATAGGATTCCC
         *   *** *   *   * ***   *       *         *   **** *    ** * * * *
```

```
T39   CTTTACATTTATTAC-GATTCTTTTTTCACGAATATTATAATTGGAATAATCTCAT---T
T41   CTTTACATTTATCACTGATTCTTTTTTCACGAATATTATAATTGGAATAATCTCAT---T
T37   CTTTACATTTATTAT-GATTCTTTTTTCACGAATATTATAATTGGAATAATCTCAT---T
T43   CTTTACATTTACTTTCTAATCTTTTTTCACGAATATTATAATTGGAATAATCTCAT---T
T38   CGATACGTTACAAAATTTAGCTTTTGACAATGATCCAATCATTGGAATAATTGGAACTAT
T40   CGATACGTTACAAAATTTAGCTTTTGACAATGATCCAATCATTGGAATAATTGGAACTAT
T42   CGATACGTTACAAAATTTAGCTTTTGACAATGATCCAATCATTGGAATAATTGGAACTAT
      *  *** **          *****  **     **    ** **********   *     *


T39   ACTCCAAAGAAATCTAACTATTATGGGTTTTCGAAAG--AGAATCCAAGACTT---TTTT
T41   ACTCCAAAGAAATCTAACTATTATGGGTTTTCGAAAG--AGAATCCAAGACTT---TTTT
T37   ACTCCAAAGAAATCTAACTATTATGGGTTTTCGAAAG--AGAATCCAAGACTT---TTTT
T43   ACTCCAAAGAAATCTAACTATTATGGGTTTTCGAAATT-AGAATCCAAGACTT---TTTT
T38   AGTTTCGAATTTTTTAGTAACAGTATCTATTAAAAAAGAACTCTCTAGCATTTGACTCTT
T40   AGTTTCGAATTTTTTAGTAACAGTATCTATTAAAAAAGAACTCTCTAGCATTTGACTCTT
T42   AGTTTCGAATTTTTTAGTAACAGTATCTATTAAAAAAGAACTCTCTAGCATTTGACTCTT
      * *     *     * **    *     * ** ***    **  *   * **    * **


T39   TGTTCCTATATAATTCTT--ATGTAGTTGAATGCGAATCCATAT--TAGTTTTTCTCCGT
T41   TGTTCCTATATAATTCTT--ATGTAGTTGAATGCGAATCCATAT--TAGTTTTTCTCCGT
T37   TGTTCCTATATAATTCTT--ATGTAGTTGAATGCGAATCCATAT--TAGTTTTTCTCCGT
T43   TGTTCCTATATAATTCTT--ATGTAGTTGAATGCGAATCCATAT--TAGTTTTTCTCCGT
T38   TACCGCTGAAGGATTTATTGGTACACTTGAAAGATAACCCAGAAAATAGAAAGAAAAATT
T40   TACCGCTGAAGGATTTATTGGTACACTTGAAAGATAACCCAGAAAATAGAAAGAAAAATT
T42   TACCGCTGAAGGATTTATTGGTACACTTGAAAGATAACCCAGAAAATAGAAAGAAAAATT
      *     **  *   ***  *    *    * ***** *   ** *** *    ***      *


T39   AAACAATCCTCTTATTTACAATCAACATCTTCTGGAACCTTTCTTGAGCGAACACATTTC
T41   AAACAATCCTCTTATTTACAATCAACATCTTCTGGAACCTTTCTTGAGCGAACACATTTC
T37   AAACAATCCTCTTATTTACAATCAACATCTTCTGGAACCTTTCTTGAGCGAACACATTTC
T43   AAACAATCCTCTTATTTACAATCAACATCTTCTGGAACCTTTCTTGAGCGAACACATTTC
T38   GGAGAATTGGCTTATGTGGATCCTAC-TGGGTTGAGACCAAAAGTGAA-AATGACATTGC
T40   GGAGAATTGGCTTATGTGGATCCTAC-TGGGTTGAGACCAAAAGTGAA-AATGACATTGC
T42   GGAGAATTGGCTTATGTGGATCCTAC-TGGGTTGAGACCAAAAGTGAA-AATGACATTGC
      *  ***    *****  *   *   * ** *   **  ***    *** *   *****  *


T39   TATGAAAAAATAGAACAACATCTCGTAGTACTTTGTTGTAA--TGATTTGCAGAAAACCC
T41   TATGAAAAAATAGAACAACATCTCGTAGTACTTTGTTGTAA--TGATTTGCAGAAAACCC
T37   TATGAAAAAATAGAACAACATCTCGTAGTACTTTGTTGTAA--TGATTTGCAGAAAACCC
T43   TATGAAAAAATAGAACAACATCTCGTAGTACTTTGTTGTAA--TGATTTGCAGAAAACCC
T38   CAAAAATTGACAAAGTAAGATTTC-CATTTCTTCATCAGAAGATGAGTCCCTTTTGAAGC
T40   CAAAAATTGACAAAGTAAGATTTC-CATTTCTTCATCAGAAGATGAGTCCCTTTTGAAGC
T42   CAAAAATTGACAAAGTAAGATTTC-CATTTCTTCATCAGAAGATGAGTCCCTTTTGAAGC
      *  **    * * *   ** ** **   * * ***  *    **  *** *     *   *


T39   TATGGTTGTTCAAGGATCCTTTCATACATTATGTTAGATATCAAGGAAAATCAATTCTGG
T41   TATGGTTGTTCAAGGATCCTTTCATACATTATGTTAGATATCAAGGAAAATCAATTCTGG
T37   TATGGTTGTTCAAGGATCCTTTCATACATTATGTTAGATATCAAGGAAAATCAATTCTGG
T43   TATGGTTGTTCAAGGATCCTTTCATACATTATGTTAGATATCAAGGAAAATCAATTCTGG
T38   CAGAATTGATTTTCCTTGATATCTAACATAATGTATGAAAGGATCCTTGAACAACCATAG
T40   CAGAATTGATTTTCCTTGATATCTAACATAATGTATGAAAGGATCCTTGAACAACCATAG
T42   CAGAATTGATTTTCCTTGATATCTAACATAATGTATGAAAGGATCCTTGAACAACCATAG
      * *** *      *  * * ** ** **  * * **** ****  ** *     * *** * *
```

```
T39    CTTCAAAAGGGACTCATCTTCTGATGAAGAAATG-GAAATCTTACTTTGTCAATTTTTGG
T41    CTTCAAAAGGGACTCATCTTCTGATGAAGAAATG-GAAATCTTACTTTGTCAATTTTTGG
T37    CTTCAAAAGGGACTCATCTTCTGATGAAGAAATG-GAAATCTTACTTTGTCAATTTTTGG
T43    CTTCAAAAGGGACTCATCTTCTGATGAAGAAATG-GAAATCTTACTTTGTCAATTTTTGG
T38    GGTTTTCTGCAAATCAT--TACAACAAAGTACTACGAGATGTTGTTCTATTTTTTCATAG
T40    GGTTTTCTGCAAATCAT--TACAACAAAGTACTACGAGATGTTGTTCTATTTTTTCATAG
T42    GGTTTTCTGCAAATCAT--TACAACAAAGTACTACGAGATGTTGTTCTATTTTTTCATAG
       *     *  * ****  *    *   *** *  *   ** ** **   * * *   ** * *

T39    CAATGTCATT-TTCACTTTTGGTCTCAACCCA-GTAGGATCCACATAAGCCAATTCTCCA
T41    CAATGTCATT-TTCACTTTTGGTCTCAACCCA-GTAGGATCCACATAAGCCAATTCTCCA
T37    CAATGTCATT-TTCACTTTTGGTCTCAACCCA-GTAGGATCCACATAAGCCAATTCTCCA
T43    CAATGTCATT-TTCACTTTTGGTCTCAACCCA-GTAGGATCCACATAAGCCAAGTCTCCA
T38    AAATGTGTTCGCTCAAGAAAGGTTCCAGAAGATGTTGATTGTAAATAAGAGGATTGTTTA
T40    AAATGTGTTCGCTCAAGAAAGGTTCCAGAAGATGTTGATTGTAAATAAGAGGATTGTTTA
T42    AAATGTGTTCGCTCAAGAAAGGTTCCAGAAGATGTTGATTGTAAATAAGAGGATTGTTTA
       * **** *   *** *     *** **    * **   *   * ***** * * * *

T39    ATTTTTCTTTCTATTTTCTGGGTTATCTTTCAAGTGTACCAATAAATCCTTCAGCGGTAA
T41    ATTTTTCTTTCTATTTTCTGGGTTATCTTTCAAGTGTACCAATAAATCCTTCAGCGGTAA
T37    ATTTTTCTTTCTATTTTCTGGGTTATCTTTCAAGTGTACCAATAAATCCTTCAGCGGTAA
T43    ATTATTCTTTCTATTTTCTGGGTTATCTTTCAAGTGTACCAATAAATCCTTCAGCGGTAA
T38    CGGAGAAAAACTAATA--TGGATTCGCATTCAACTACATAAG--AATTATATAGGAACAA
T40    CGGAGAAAAACTAATA--TGGATTCGCATTCAACTACATAAG--AATTATATAGGAACAA
T42    CGGAGAAAAACTAATA--TGGATTCGCATTCAACTACATAAG--AATTATATAGGAACAA
       *** *   *** **  * *****  *  *   *   ** *  *   **   **

T39    AGAGTCAAATGCTAGAGAGTTCTTTTTTAATAGATACTGTTACTAAAAAATTCGAAACTA
T41    AGAGTCAAATGCTAGAGAGTTCTTTTTTAATAGATACTGTTACTAAAAAATTCGAAACTA
T37    AGAGTCAAATGCTAGAGAGTTCTTTTTTAATAGATACTGTTACTAAAAAATTCGAAACTA
T43    AGAGTCAAATGCTAGAGAGTTCATTTTTAATAGATACTGTTACTAAAAAATTCGACACTA
T38    AAA---AAGTCTTGGA---TTCTCTTTCGAAAACCCATAATAGTTAGAT-TTCTTTGGAG
T40     AAA---AAGTCTTGGA---TTCTCTTTCGAAAACCCATAATAGTTAGAT-TTCTTTGGAG
T42     AAA---AAGTCTTGGA---TTCTCTTTCGAAAACCCATAATAGTTAGAT-TTCTTTGGAG
       * *    ** *  * *  **    *** *** ***  * *      *   ** * * *    ***

T39    TAGTTCCAATTATTCCAATGATTGGATCATTGTCAAAAGCTAAATTTTGTAA-CGTATCG
T41    TAGTTCCAATTATTCCAATGATTGGATCATTGTCAAAAGCTAAATTTTGTAA-CGTATCG
T37    TAGTTCCAATTATTCCAATGATTGGATCATTGTCAAAAGCTAAATTTTGTAA-CGTATCG
T43    TAGTTCCAATTATTTCAATGATTGGATCATTGTCAAAAGCTAAATTTTGTAA-CGTATCG
T38    TAATGA-GATTATTCCAATTAT--AATATTCGTGAAAAAGAATCGTAATAAATGTAAAG
T40    TAATGA-GATTATTCCAATTAT--AATATTCGTGAAAAAGAATCGTAATAAATGTAAAG
T42    TAATGA-GATTATTCCAATTAT--AATATTCGTGAAAAAGAATCGTAATAAATGTAAAG
       ** *     ****** **** **   **  * ** ****   **   *  ***  *** *

T39    GGGAATCC-TATTAGTAAGCCAGTTTGGACCGATTTGTCAGATTCTGATATTATTGATCG
T41    GGGAATCC-TATTAGTAAGCCAGTTTGGACCGATTTGTCAGATTCTGATATTATTGATCG
T37    GGGAATCC-TATTAGTAAGCCAGTTTGGACCGATTTGTCAGATTCTGATATTATTGATCG
T43    CGGAATCCATATTAGTAAGCCAGTTTGGACCGACTTGTCAGATTCTGATATTATTGATCG
T38    AAGGAACATCTTGTATCCAGCATTGTAGAATTTGAACCAAGATTTCTAGATGTACGGGAT
T40    AAGGAACATCTTGTATCCAGCATTGTAGAATTTGAACCAAGATTTCTAGATGTACGGGAT
T42    AAGGAACATCTTGTATCCAGCATTGTAGAATTTGAACCAAGATTTCTAGATGTACGGGAT
       * *  *     *     *     *   ** * *  **      *****   * **     *
```

```
T39    ATTTGGTCGG-ATATGTAGAAATCTTT-CTCATTATTACAGT-GGGTCTTCAAAAAAACA
T41    ATTTGGTCGG-ATATGTAGAAATCTTT-CTCATTATTACAGT-GGGTCTTCAAAAAAACA
T37    ATTTGGTCGG-ATATGTAGAAATCTTT-CTCATTATTACAGT-GGGTCTTCAAAAAAACA
T43    ATTTGGTCGGGATATGTAGAAATCTTTTCTCATTATTACAGTTGGGTCTTCAAAATAACA
T38    AGGGTATTAGTATATCTGATACACAAT-TTAAATGTGATAAT-TTGTCCTCAAAATGA-G
T40    AGGGTATTAGTATATCTGATACACAAT-TTAAATGTGATAAT-TTGTCCTCAAAATGAAG
T42    AGGGTATTAGTATATCTGATACACAAT-TTAAATGTGATAAT-TTGTCCTCAAA-TGAGT
       *     *  * **** *     *     *    *    * * * * * *   *** *****   *


T39    AAGTTTGTATCGAATAAAGTAT
T41    AAGTTTGTATCGAATAAAGTAT
T37    AAGTTTGTATCGAATAAAGTAT
T43    AAGATAGTATCGAATAAAGTAT
T38    TGAATAAAAATTGGGGAGG---
T40    TAAAAAAAAAAGAAGAGG---
T42    TTAATATTAGGGGGGGGGG---
          *          *
```

Annex VI (Alignment of the *psbA-trnH* intergenic region sequences)

```
CLUSTAL 2.1 multiple sequence alignment
T37    --------------------ATATAATATTATATATATATAGAATATTATAATATTTC
T39    ------ATTCTAAGAATATATATATAATATTATATATATATAGAATATTATAATATTTC
T38    ------ATATATATATATATATATATAATATTATATATATATAGAATATTATAATATTTC
T43    ATATATATATATATATATATATATAGAATATTATATATATATAGAATATTATAATATTTC
T41    --------------------ACTTGGCTACATCCGCCAGAACCACTTGGCTACTTC
T40    ------ATATATATATATATATATATAATATTATATATATATAGAATATTATAATATTTC
T42    ------ATATATATATATATATATATAATATTATATATATATAGAATATTATAATATTTC
                            *      ** **           *    *   *     ** ***


T37    TATATTTATTATATA-ATATTTCTATTATTATTTCTTTTTTATATTAATATTCTATTTTA
T39    TATATTTATTATATA-ATATTTCTATTATTATTTCTTTTTTATATTAATATTCTATTTTA
T38    TATATTTATTATATA-ATATTTCTATTATTATTTCTTTTTTATATTAATATTCTATTTTA
T43    TATATTTATTATATA-ATATTTCTATTATTATTTCTTTTTTATATTAATATTCTATTTTA
T41    CGCCTGATCCACTTGGATACATCCGCC--TGATTCTTTTTTATATAAATATTCTATTTTA
T40    TATATTTATTCTATA-ATATTTCTATTATTATTTCTTTTTTATATTAATATTCTATTTTA
T42    TATATTTATTATATA-ATATTTCTATTATTATTTCTTTTTTATATTAATATTCTATTTTA
       *         *   *** **        *   ************* ***************


T37    TTTATTTATATTTATTTATTATTTCTATTTTTATTTTTTATATATTTATTTTTCTTTTTG
T39    TTTATTTATATTTATTTATTATTTCTATTTTTATTTTTTATATATTTATTTTTCTTTTTG
T38    TTTATTTATATTTATTTATTATTTCTATTTTTATTTTTTATATATTTATTTTTCTTTTTG
T43    TTTATTTATATTTATTTATTATTTCTATTTTTATTTTTTATATATTTATTTTTCTTTTTG
T41    TTTATTTATATTTATTTATTATTTCTATTTTTATTTTTTATATATTTATTTTTCTTTTTG
T40    TTTATTTATATTTATTTATTATTTCTATTTTTATTTTTTATATATTTATTTTTCTTTTTG
T42    TTTATTTATATTTATTTATTATTTCTATTTTTATTTTTTATATATTTATTTTTCTTTTTG
       ***********************************************************


T37    TTTTATATTTAACTTTTTTATTCTTATTCTTTGCTTTGTATTTTTTTTGTATTTAGTTTT
T39    TTTTATATTTAACTTTTTTATTCTTATTCTTTGCTTTGTATTTTTTTTGTATTTAGTTTT
T38    TTTTATATTTAACTTTTTTATTCTTATTCTTTGCTTTGTATTTTTTTTGTATTTAGTTTT
T43    TTTTATATTTAACTTTTTTATTCTTATTCTTTGCTTTGTATTTTTTTTGTATTTAGTTTT
T41    TTTTATATTTAACTTTTTTATTCTTATTCTTTGCTTTGTATTTTTTTTGTATTTAGTTTT
T40    TTTTATATTTAACTTTTTTATTCTTATTCTTTGCTTTGTATTTTTTTTGTATTTAGTTTT
T42    TTTTATATTTAACTTTTTTATTCTTATTCTTTGCTTTGTATTTTTTTTGTATTTAGTTTT
       ***********************************************************
```

```
T37    TCGTTCATACATTTATTTTGTATTTCAAATAAATAAAAGATTTGATTTCAAAGCCAAAGA
T39    TCGTTCATACATTTATTTTGTATTTCAAATAAATAAAAGATTTGATTTCAAAGCCAAAGA
T38    TCGTTCATACATTTATTTTGTATTTCAAATAAATAAAAGATTTGATTTCAAAGCCAAAGA
T43    TCGTTCATACATTTATTTTGTATTTCAAATAAATAAAAGATTTGATTTCAAAGCCAAAGA
T41    TCGTTCATACATTTATTTTGTATTTCAAATAAATAAAAGATTTGATTTCAAAGCCAAAGA
T40    TCGTTCATACATTTATTTTGTATTTCAAATAAATAAAAGATTTGATTTCAAAGCCAAAGA
T42    TCGTTCATACATTTATTTTGTATTTCAAATAAATAAAAGATTTGATTTCAAAGCCAAAGA
       ************************************************************

T37    AGTACATAATTACGGATTGGTAAAAGCCAAAGTATGCTAAAATATTAAAATAAAAGTACA
T39    AGTACATAATTACGGATTGGTAAAATCCTAAGTATGCTAAAATATTAAAATAAAAGTACA
T38    AGTACATAATTACGGATTGGTAAAAGCCAAAGTATGCTAAAATATGAAAATAAAAGTACA
T43    AGTACATAATTACGGATTGGTAAAAGCCAAAGTATGCTAAAATATTAAAATAAAAGTACA
T41    AGTACATAATTACGGATTGGTAAAAGCCAAAGTATGCTAAAATATTAAAATAAAAGTACA
T40    AGTACATAATTACGGATTGGTAAAAGCCAAAGTATGCTAAAA-ATAAAAATAAAAGTACA
T42    AGTACATAATTACGGATTGGTAAAAGCCAAAGTATGCTAAAATATTAAAATTAAAGTTCA
       ************************** ** ************* ** ***** ***** **

T37    AATGTCAAAAATGCTTATGTGGACAAAATCCCAGTGAATCAAAAAAAGGAGTAATACCAA
T39    AATGTCAAAAATGCTTATGTGGACAAAATCCCAGTGAATCAAAAAAAGGAGTAATACCAC
T38    AATGTCAAAAATGCTTATGTGGACAAAATCCCAGTGAATCAAAAAAAGGAGTAATACCCA
T43    AATGTCAAAAATGCTTATGTGGACAAAATCCCAGTGAATCAAAAAAAGGAGTAATACCAA
T41    AATGTCAAAAATGCTTATGTGGACAAAATCCCAGTGAATCAAAAAAAGGAGTAATAC-CA
T40    AATGTCAAAAACGCTTATGTGGACA-AAATTCCCCACTTCTATACAT----CCGCGTGAT
T42    AATGTCAAAAATAGTTATTTGAACACAAGGTTAGCACTTCCATAC------CAAGGTTAG
       ***********    **** ** *** **          ** *  *

T37    -ACCTCCTCATTAGAGGTGTGGTATTATTCCTTCAACAATTCCTATACACTAAGACAAAA
T39    -ACCTCCTCATTAGAGGTTTGGTATTATTCCTTCAACAATTCCTATACACTAAGACAAAA
T38    TAGTAGGGCATTAGAGGTTTGGTATTATTCCTTCAACAATTCCTAGACTCAAAGACAAAA
T43    -ACCTCCTCATTAGAGGTTTGGTATTATTCCTTCAACAATTCCTATACGAGAAGACAAAA
T41    AACCTCCTCATTAGAGGTTTGGTATTATTCCTTCAACAATTCCTATACACTAAGACAAAA
T40    CATCTCGTAGGCTCCGCT-TG--ATCCACTTGGCTACATCCGCTTGATCCACTTACTAGA
T42    CACTTC-CAAACCAAGGT-TAGCACTTCCATACCAAGGTTAGCACTTCCATACCATTAAG
         *        * * *    *          * *         *          *  *

T37    TGTCTTATCCATTTGTAGATGGAACTTCAACAGCAG--CTAAGTCTAG-AGGGAAGTTGT
T39    TGTCTTATCCATTTGTAGATGGAACTTCAACAGCAG--CTAAGTCTAG-AGGGAAGTTGT
T38    TGTCTTATCCATTTGTAGATGGAACTTCAACAGCAG--CTAAGTCTAG-AGGAACGAGA-
T43    TGTCTTATCCATTTGTAGATGGAACTTCAACAGCAG--CTAAGTCTAG-ATGTAAGCTG-
T41    TGTCTTATCCATTTGTAGATGGAACTTCAACAGCAG--CTAAGTCTAG-AGGGAAGTTGT
T40    TCCGCTTGATCCACTTGGCTACATCCGCCTGATCCA--CTGGCTACACCGCCTGTACA-T
T42    GGCACTTCCATACCAAGGTTAGCACTTCCATACCAAGGTTAGCACTTCCATCCAAGTTCT
          *            * *   *   *   * *        *

T37    GAGCATTACGTTCATGCATTACTTCCATCCGAGTGTT----------------------
T39    GAGCATTACGTTCATGCATTACTTCCATCCGGGGGGA----------------------
T38    -A--CTTAAATCGGA--------------------------------------------
T43    -ATCCTTGTACCG----------------------------------------------
T41    GAGCATTACGTTCATGCATTACTTCCTCCAAGGGGTTAGACACACCCCCCCCCCAGAGGA
T40    GATATTTTCCTCGGC--------------------------------------------
T42    AAAACCTTCCTGGGT--------------------------------------------
         *    *
```

Toxic Elements in Rice and Possible Health Risk Assessment – Bangladesh Prospect

Yeasmin Nahar Jolly[1], Roksana Huque[2], Ashraful Islam[1],
Safiur Rahman[1], Shirin Akter[1], Jamiul Kabir[1],
Kamruzzaman Munshi[2], Mahfuza Islam[2], Afifa Khatun[2], Arzina Hossain[2]

1 Introduction

Heavy metals are highly persistent and non-biodegradable contaminants that cause toxic effect in humans and may bio-accumulated through food chain to hazardous level, thus posing potential health risks to human by consumption.The sources of heavy metal in plants are their growth media like air, soil, nutrients etc. from which heavy metals are taken up by roots or foliage. Cu, Zn, Mn, and Fe are essential heavy metals in plants nutrition but many heavy metals do not play any significant role in the plants physiology.

Anthropogenic activities like industrialization have being contributing to the spread of toxic chemicals into the environment, increasing the levels of human exposure to many of them(Rahman *et al.*, 2014). In Bangladesh almost all industries are seen to discharge their wastes into water and on land without any treatment or after partial treatment. Toxic elements from these wastes may contaminate agricultural soils, water supplies and environment and hence food chain. As, Cd, Hg, and Pb are the ubiquitous trace elements known to have a harmful effect on human health (Rahman *et al.*, 2012). These elements are naturally present at very low concentrations in environment, and human bodies are able to detoxify them in limited amounts. Plants growing in a polluted envi-

[1] Chemistry Division, Bangladesh Atomic Energy Commission, Bangladesh
[2] Food Technology Division, Bangladesh Atomic Energy Commission, Bangladesh

ronment can accumulate the toxic metals at high concentration causing serious risk to human health when consumed. Food is the most important route for accumulating most chemical (essential and toxic). In a study Wagner (1993) reported that more than 70% of dietary intake of cadmium is contributed via food chain (Chaney 1980; Smith *et al.*, 1996) cautioned on the use of waste water in crop production since it may be possible for heavy metal from the waste to accumulate in soils and thereby enter the food chain, contaminate surface and underground water thus cause health hazard. In another study Saleemi (1990) reported that serious health problems can develop as a result of accumulation of dietary heavy metal uptake through food crops irrigated with contaminated water.

Rice (*Oryza sativa*) is the staple food of about 135 million people of Bangladesh. It provides nearly 48% of rural employment, about two-third of total calorie supply and about one-half of the total protein intakes of an average person in the country. Rice sector contributes one-half of the agricultural GDP and one-sixth of the national income in Bangladesh. About 75% of the total cropped area and over 80% of the total irrigated area is planted to rice. Thus, rice plays a vital role in the livelihood of the people of Bangladesh. Recently, concern has been raised about possible contamination of rice worldwide by heavy metals (Fu *et al.*, 2008; Zhao *et al.*, 2010), which is more acute in case of Bangladesh. The present study was conducted to assess the heavy metal (Cr, Co, Ni, Cd, Pb, As, Cu, Zn, and Mn) concentration in soils of rice field, irrigated water and rice grown in different region of Bangladesh. Transfer of heavy metal from soil to rice and assessment of potential health risk of those metals to human by consumption of rice was also calculated accordingly.

2   Material and Methods

2.1   Site Description

A map of sampling sites is shown in Figure 1. Samples were collected from eight different districts of Bangladesh with GPS (Geographical Position System) reading (Table 1). These were Joypurhat, Naogaon, Shatkhira, Jessore, Khagrachari, Bandarban, Rangamati and Savar.

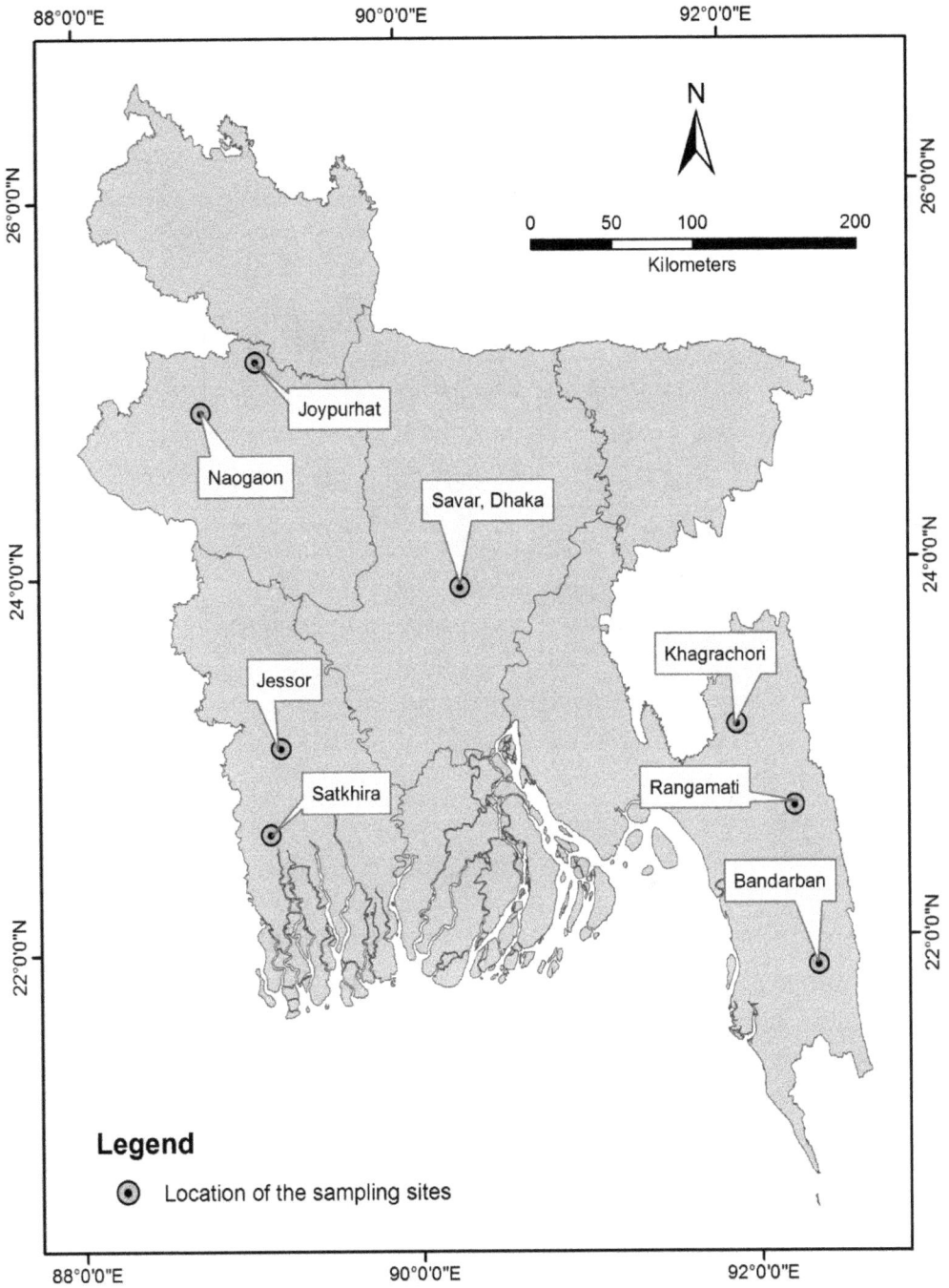

Figure 1: Location maps of sampling sites.

2.2   Collection of Samples

Fresh polyethylene bottles with double stoppers were used for collection of water samples. The solid samples were collected in newly purchased fresh small polyethylene bags. The sample bottles after treatment with detergent solution were washed with tap water, kept immersed in 5% HNO_3 acid water overnight,

Areas	Latitude	Longitude
Joypurhat	25.1000° N	89.1000° E
Naogaon	24.9000° N	88.7500° E
Satkhira	22.3500° N	89.0800° E
Jessor	23.1700° N	89.2000° E
Khagrachori	23.0417° N	91.9944° E
Bandarban	21.8000° N	92.4000° E
Rangamati	22.6333° N	92.2000° E
Savar	23.8583° N	90.2667° E

Table1: Latitude and longitude of sampling areas

rinsed with sufficient amount of deionized water and finally dried in air. The bottles and the bags were marked with special identification numbers according to the sample types.

The irrigation water has been collected in polyethylene bottles from the rice field of different location. The rice samples (Aman- non aromatic, 24 samples) were collected in clean polyethylene bags from three different rice field for each sampling sites and remove the husk and dried for 24hr in an oven at 35ºC in a glass beaker to constant weight. Then the rice samples were ground to fine powder by using a blender and finally with mortar and pestle, preserved in a desiccators until analysis. The soil samples have been collected from three different rice fields for each sampling sites (from where the rice samples had been collected) at a depth of 0 – 25 cm and carried in thick polyethylene bags. The soil samples were oven dried (35 ºC), sieved and crushed homogeneously and preserved in a desiccators until further processing. A minimum of 3 pellets of each samples (soil and rice) were made to reduce the error in the analysis.

2.3   Sample Preparation

2.3.1 Water Sample

A volume of 500 mL of each collected water sample filtered with Whatman 41 filter paper was taken in a clean weighed porcelain dish followed by addition of 4gmof cellulose powder and evaporated on water bath. The sample after evaporation to dry mass was further dried under IR lamp at about 70 ºC for two hours to remove the trace of moisture and weighed. For homogeneous mixing, the dry mass was then transferred to a carbide mortar and ground to fine powder using a pestle. The processed sample in a plastic vial with identification mark was preserved inside desiccators. A minimum of 3 pellets of each samples were made to reduce the error in the analysis.

2.3.2 Soil Sample

Collected soil samples were further dried at 60 °C in an oven until constant weight was obtained. The dried samples were ground to fine power in an Agate mortar with a pestle and preserved in polyethylene bags in a desiccator for further analysis (Jolly *et al.*, 2013b). A minimum of 3 pellets of each samples were made to reduce the error in the analysis.

2.3.3 Rice Sample

The rice samples were further dried in an oven at 60 °C until constant weight was obtained. The dried samples were finally ground in a carbide mortar with a pestle and preserved in polyethylene bags in a desiccator until subsequent analysis. A minimum of 3 pellets of each samples were made to reduce the error in the analysis.

2.4   Analysis of Heavy Metal

The Panalytical Epsilon 5 Energy Dispersive X-ray Fluorescence (EDXRF) was used as major analytical technique for carrying out elemental analysis in the samples (water, soil and rice). The method was described in detailed elsewhere

(Jolly *et al.*, 2013a). For irradiation of the sample with X-ray beam 2 g of each powdered material was pressed into a pellet of 25 mm diameter with a pellet maker (Specac) and loaded into the X-ray excitation chamber with the help of automatic sample changer system. A time-based programme, controlled by a software package provided with the system was assigned for the irradiation of all real samples as well as the standards. The X-ray intensities of the elements in sample spectrum were calculated using the system software by integration of area of the respective X-ray peak areas using peak fitting deconvolution software.

2.4.1 Quality Control Analysis

Deionized water and 10% HNO_3 was used for glassware washing. Precision and accuracy of the method was checked through repeated analysis of Montana-1/2710a for soil, Spinach/NIST 1570a for rice and NIST water SRM 1640 for water for heavy metals. Obtained values were found in good agreement with the certified values and the percentage of relative error and coefficient of variation in almost all the elements were less than 10% (Jolly *et al.*, 2013a; Jolly *et al.*, 2013b).

2.4.2 Data Analysis

Transfer Factor

Soil to rice metal transfer was computed as transfer factor (TF), which was calculated by using the following equation (Eq. (1)).

$$TF = C_{plant} / C_{Soil} . \tag{1}$$

where C_{plant} and C_{soil} represents the toxic metal concentration in extracts of rice and soils on dry weight basis, respectively.

Metal Pollution Index (MPI)

Metal pollution index (MPI) was computed to determine overall heavy metal concentration in each varieties of rice samples analysed. This index was obtained by calculating the geometrical mean of concentrations of all the metals in rice sample collected (Ureso *et al.*, 1997) using the following equation (Eq. (2)).

$$\text{MPI } (\mu g\ g^{-1}) = (Cf_1 \times Cf_2 \times \ldots \times Cf_n)^{1/n}. \tag{2}$$

where Cf_n= concentration of metal in n in the sample

Daily Intake of Metals (DIM)

Daily intake was calculated by the following equation (Eq. (3)).

$$\text{M Daily intake of metal (DIM)} = \frac{C_{metal} \times D_{food\ intake}}{B_{average\ weight}}. \tag{3}$$

where C_{metal}, $D_{food\ intake}$, and $B_{average\ weight}$ represent the heavy metal concentrations in rice ($\mu g\ g^{-1}$), daily intake of rice and average body weight, respectively. The FAO statistics show that each person in Bangladesh consumes 160 kilograms of rice a year, so from that statistics, daily consumption per person is 438gm (approximately)(Daily Star, December. 12, 2011). The average body weight ($B_{average\ weight}$) was taken as 70 kg for adults according to WHO guideline (WHO, 1993).

Health Risk Index (HRI)

Value of Health Risk Index (HRI) depends on the daily intake of metals (DIM) through foodstuff and oral reference dose (RfD). RfD is an estimated per day exposure of metal to human body that has no hazardous effect during life time (US-EPA IRIS, 2006). The health risk index for Cr, Co, Ni, Cu, Pb, Cd, Mn and Zn by consumption of rice was calculated by the following equation (Eq. (4)) obtained from literature (Cui *et al.*, 2004).

$$\text{HRI} = \text{DIM} / \text{RfD}. \tag{4}$$

Oral reference doses For Cr, Ni, Cu, Pb, Cd, Mn and Zn were 1.5, 0.02, 0.04, 0.004, 0.001, 0.033 and 0.30 (mg/kg bw/day) respectively (US-EPA IRIS, 2006) and 3.01 mg day^{-1} for Co (Food and Nutritional Board, 2004). The Reference Dose (RfD) for inorganic arsenic is 0.0003 mg day^{-1} based on hyper pigmentation, keratosis, and possible vascular complications in humans (US EPA, 1998). Estimated exposure is obtained by dividing daily intake of heavy metals by their safe limit. An index more than 1 is considered as not safe for human health (USEPA, 2002).

Hazard Index (HI)

To evaluate the potential risk to human health through more than one heavy metal, the hazard index (HI) has been developed (USEPA, 1989). The hazard index is the sum of the hazard quotients as described in the following equation (Eq. (5)).

$$HI = \sum HQ = HQ_{Cr} + HQ_{Co} + HQ_{Cd} + HQ_{Pb} + HQ_{As}. \tag{5}$$

It is assume that magnitude of adverse effect will be proportional to the sum of multiple metal exposures.

2.5   Statistical Analysis

Statistical procedures were performed using SPSS for Microsoft version 17.0 software package (SPSS Chicago, IL). One way ANOVA was applied for evaluating the significant different between heavy metal concentration in Soil, water and rice sample of different districts.

3   Results and Discussion

3.1   Concentration of Heavy Metals in Water Collected from Rice Field

Concentration of different heavy metal analyzed with reference is shown in Table 2. ANOVA showed that irrigated water from Bandarban and Rangamati area showed significantly highest concentration of chromium compare to other areas. There was no significant difference in chromium level in irrigation water among Khagrachori, Jessor and Savar. Highest level of cobalt was found in irrigation water from Joypurhat while lowest level of cobalt was determined in water from Khagrachori. Maximum level of nickel was found in Rangamati (0.045 mg/L) followed by Bandarban (0.038 mg/L). Irrigation water from Shatkhira had maximum copper concentration 0.24 mg/L compare to other locations. On the other hand Joypurhat, Naogaon, Khagrachori, Jessor and Savar showed more or less same amount of copper. A significantly ($P < 0.001$) maximum arsenic level (0.021 mg/L) was found in irrigation water from Shatkhira, while Savar showed minimum amount of arsenic (0.0087 mg/L). Arsenic level was not significantly

Sample Id	Element (Mean ± SD) mg/L								
	Cr	Co	Ni	Cu	As	Pb	Cd	Zn	Mn
Joypurhat	< 0.001	0.079 ± 0.002a	0.018 ± 0.002a	0.019 ± 0.002a	0.01 ± 0.001a	0.017 ± 0.001a	0.0063 ± 0.001a	0.04 ± 0.001a	0.25 ± 0.01a
Naogoan	< 0.001	0.061 ± 0.003ab	0.015 ± 0.001b	0.017 ± 0.001ab	0.01 ± 0.001ab	0.018 ± 0.002ab	7.87E-05 ± 0.0001b	0.032 ± 0.001ab	0.34 ± 0.02b
Shatkhira	< 0.001	< 0.001	0.024 ± 0.001c	0.24 ± 0.014c	0.021 ± 0.001c	0.022 ± 0.01bc	0.005 ± 0.001ac	0.016 ± 0.001c	0.087 ± 0.002c
Khagrachori	0.025 ± 0.005a	0.006 ± 0.001c	0.02 ± 0.001d	0.02 ± 0.001abd	0.01 ± 0.001d	0.015 ± 0.001abd	< 0.0017	0.20 ± 0.012d	0.87 ± 0.08d
Rangamati	0.04 ± 0.001b	< 0.001	0.045 ± 0.001e	0.044 ± 0.001e	< 0.004	0.03 ± 0.001e	< 0.0017	0.089 ± 0.005e	0.035 ± 0.005e
Bandarban	0.05 ± 0.007bc	< 0.001	0.038 ± 0.001f	0.045 ± 0.001ef	< 0.004	0.03 ± 0.002ef	< 0.0017	0.1 ± 0.001f	0.03 ± 0.001f
Jessore	0.02 ± 0.004ad	< 0.001	0.02 ± 0.001ag	0.02 ± 0.001abdg	0.014 ± 0.002e	0.014 ± 0.002adg	< 0.0017	0.013 ± 0.002cg	0.18 ± 0.006g
Savar	0.028 ± 0.01ade	< 0.001	0.02 ± 0.001dh	0.02 ± 0.001abdgh	0.0087 ± 0.001f	0.014 ± 0.001abdgh	< 0.0017	0.031 ± 0.002abh	0.11 ± 0.01ch
DoE Standardx	1.0	-	1.0	3.0	0.2	0.1	0.5	10.0	5.0

Table 2: Concentration of different heavy metals in water samples collected from different rice fields. x = Department of Environment (DoE) (1999). Values are mean of 3 replicates. In this table, mean values with different superscript letters are significantly different at $P = 0.05$ and LSD values are at $P = 0.05$.

different between Joypurhat and Naogaon. Lead has been determined in highest amount in irrigation water from Rangamati and Bandarban rice field than other rice field but there was no significant difference in lead level between these two areas. Statistical analysis showed that there was no significant difference in cadmium level between irrigation water from Joypurhat and Shathkhira. Cadmium level was significantly different between Joypurhat and Naogaon. ANOVA showed that a significantly ($P < 0.001$) highest concentration of zinc was found in irrigation water from Khagrachori followed by Bandarban and Rangamati. Zinc level was not significantly different among Joypurhat, Naogaon and Savar. Highest level of manganese was determined in water from Khagrachori compare to other locations. Lower limit of manganese was found in irrigation water from Bandarban and Rangamati. There was no significant difference in manganese level between Shatkhira and Savar.

3.2   Concentration of Heavy Metals in Soil Collected from Rice Field

Heavy metal concentration (Cr, Co, Ni, Cu, As, Pb, Cd, Zn and Mn) for soil collected from different district is presented in Table 3. It was found that concentration of heavy metals in soil samples varied among the different districts. Concentrations of Cr in Joypurhat, Nowgaon, Shatkhira with an exception of Rangamati, Bandorban, Khagrachori, Jessore, Savar are within the permissible limit suggested by European standard (2006). Statistical analysis showed that a significantly ($P < 0.001$) highest level (558.25 mg/kg) of chromium was found in the soil of Savar area compare to those of other locations. The chromium level was found to be lower in the soil of Joypurhat, Naogaon and Shatkhira compare to other experimental areas. Concentration of Co, Cu, Cd, Zn and Mn are found within the permissible limit set by the standard value according to European standard. Cobalt was found in higher concentrations (27.60 mg/kg) in the soil of Rangamaticompare to other locations. Cobalt concentrations were not significantly different in the soil of Joypurhat, Naogaon, Jessor and Savar areas. There was no significant difference in cadmium level in the soil of all experimental areas. Statistical analysis for arsenic level in soil showed that there was no significant difference between Joypurhat and Naogaon with lower level of arsenic. Soil of Jessor showed a significantly highest amount (56.53 mg/kg) of arsenic compare to other areas. A significantly highest amount of copper (80.17 mg/kg)

Sample Id	Element (Mean ± SD), mg/kg								
	Cr	Co	Ni	Cu	As	Pb	Cd	Zn	Mn
Joypurhat	58.91 ±3.02[a]	13.32 ±2.09[a]	<0.19	38.33 ±2.17[abcd]	38.33 ±0.09[a]	41.2 ±0.98	0	69.88 ±0.02[a]	592.18 ±5.2[a]
Naogoan	70.82 ±0.94[ab]	12.46 ±1.0[ab]	<0.19	38.36 ±1.04[abcd]	38.53 ±0.92[ab]	<1.35	1.176 ±0.192[a]	84.34 ±1.99[b]	608.84 ±4.62[ab]
Shatkhira	67.47 ±1.01[abc]	18.16 ±1.02[c]	<0.19	-	<4.17	<1.35	1.38 ±0.14[a]	106.33 ±1.0[c]	0
Khagrachori	198.43 ±28.62[d]	19.89 ±0.45[cd]	<0.19	45.11 ±0.31[abcd]	51.22 ±4.95[c]	<1.35	1.24 ±0.54[a]	97.39 ±1.7[cd]	643.49 ±28.62[abcd]
Rangamati	110.59 ±0.26[e]	27.60 ±0.24[e]	<0.19	51.5 ±0.31[abcd]	<4.17	<1.35	1.85 ±0.68[a]	79.87 ±1.89[abef]	658.80 ±24.59[bcde]
Bandarban	156.28 ±34.2[ef]	7.66 ±0.69[f]	<0.19	59.56 ± 2.64[e]	51.79 ±0.91[cd]	<1.35	1.34 ±0.65[a]	89.79 ±18.74[bdefg]	710.19 ±17.69[deg]
Jessore	158.33 ±28.08[efg]	12.59 ±3.13[abg]	<0.19	62.14 ±14.33[ef]	56.53 ±2.43[e]	<1.35	1.62 ±0.69[a]	92.05 ±0.88[bdfg]	896.32 ±53.08[f]
Savar	558.24 ±13.87[h]	12.15 ±0.84[abgh]	<0.19	80.17 ±5.89[g]	46.14 ±0.58[f]	<1.35	1.97 ±0.98[a]	116.94 ±3.4[ch]	750.14 ±48.58[g]
Standard[x]	100	50.0	50	100	10.9	100	3.00	300	2000
WAV[y]	47	5.5	13	13	4.4	22	0.37	45	270

Table 3: Heavy metal concentration in soil samples. x = European Union Standards European Union (2006); y = World Average Value (Pendias and Pendias, 2000).Values are mean of 3 replicates. In this table, mean values with different superscript letters are significantly different at $P = 0.05$ and LSD values are at $P = 0.05$.

was found in the soil of Savar area compare to those of other areas. Lowest amount of copper was found in the soil of Joypuhat and Naogaon areas and there was no significance difference between these two areas. Soil of Savar and Shatkhira areas had a significantly ($P < 0.001$) highest zinc level than all other areas and there was no significant difference between them. Jessor showed a significantly ($P < 0.001$) highest level of manganese (896.32 mg/kg) in the soil than all other experimental locations. Lowest amount of manganese was found in the soil of Joypurhat and Naogaon area with no significant difference. On the other hand concentration of Ni and Pb are too low to be detected by the analytical system. According to world average value most of elements except Ni and Pb in all soil samples showed a higher value.

3.3   Concentration of Heavy Metals in Rice

Mean value of heavy metals analysed in Rice samples on dry weight basis collected from different districts along with the permissible limit set by FAO/WHO is presented in Table 4. ANOVA statistical analysis showed that a significantly ($P < 0.001$) highest amount of chromium was found in rice sample from Khagrachori compare to those from other locations. Bandorban and Rangamati showed no significant difference in chromium level. There was also no significant difference between Naogaon and Joypurhat. Chromium levels in case of all samples from all experimental areas were found to be greater than world standard limit of 0.05 mg/kg (FAO/WHO 1990). Nickel contents of rice sample were significantly ($P < 0.001$) different between Naogan (5.83 mg/kg) and Joypurhat (5.38 mg/kg) with the highest level compare to other areas. There was no significant difference in nickel content among Khagrachori, Rangamati, Jessor and Savar areas. The standard limit of nickel in plants recommended by WHO is 10 mg/kg and all the rice samples in the present study have nickel contents below this limit. So it is revealed that rice samples of experimental areas are safe from the hazardous effects of nickel. ANOVA analysis showed that rice from Joypurhat had significantly ($P < 0.001$) higher amount (8.28 mg/kg) of copper content compare to those from other experimental areas. Copper level was not significantly different between Bandorban and Rangamati. The observed copper levels in all samples were found to be lower than world standard limit of 10 mg/kg (FAO/WHO, 1984).

Sample Id	Element (Mean ± SD) mg/kg								
	Cr	Co	Ni	Cu	As	Pb	Cd	Zn	Mn
Joypurhat	6.21 ± 1.06a	< 0.22	5.35 ± 0.07a	8.28 ± 0.05a	< 0.01	2.84 ± 0.34a	< 0.06	27.47 ± 0.36a	24.94 ± 1.68a
Naogoan	6.37 ± 1.08ab	< 0.22	5.83 ± 0.33b	7.3 ± 0.04b	0.68 ± 0.042a	2.49 ± 0.23ab	< 0.06	21.41 ± 0.83b	20.93 ± 0.14b
Shatkhira	13.58 ± 0.19c	0.701 ± 0.08	2.48 ± 0.24c	4.99 ± 0.64 b	0.06 ± 0.01b	1.35 ± 0.03c	< 0.06	60.77 ± 0.73c	24.87 ± 0.50ac
Khagrachori	17.39 ± 10.03d	< 0.22	1.95 ± 0.22degh	5.77 ± 0.08d	0.30 ± 0.21c	< 0.12	2.83 ± 0.057ac	27.21 ± 0.06ad	32.55 ± 0.31d
Rangamati	2.06 ± 0.75efgh	< 0.22	2.02 ± 0.09degh	3.93 ± 0.04e	0.54 ± 0.01d	< 0.12	1.19 ± 0.025bce	30.89 ± 0.39e	32.98 ± 0.36de
Bandarban	1.30 ± 0.26efg	< 0.22	1.45 ± 0.05f	4.01 ± 0.18ef	0.54 ± 0.01de	< 0.12	1.32 ± 0.031abc	27.58 ± 0.28adf	16.99 ± 0.32f
Jessore	1.544 ± 0.33efg	< 0.22	2.05 ± 0.03degh	5.86 ± 0.08dg	0.44 ± 0.031cdef	< 0.12	1.37 ± 0.046abcd	16.82 ± 0.63g	8.92 ± 0.05ag
Savar	2.67 ± 0.24h	< 0.22	1.94 ± 0.06degh	7.58 ± 0.07bh	0.58 ± 0.04cdefg	< 0.12	2.89 ± 0.08acde	37.007 ± 0.39h	17.84 ± 0.48fh
FAO/WHO	-	-	-	20.0	-	0.2	0.2	5.0	-

Table 4: Concentration of heavy metals in Rice sample of different district. Values are mean of 3 replicates. In this table, mean values with different superscript letters are significantly different at $P = 0.05$ and LSD values are at $P = 0.05$.

Sample	Transfer factor (TF)								
Locations	Cr	Co	Ni	Cu	As	Pb	Cd	Zn	Mn
Joypurhat	0.105	0	0	0.216	0	0.069	0	0.393	0.044
Naogaon	0.090	0	0	0.190	0.018	0	0	0.254	0.034
Shatkhira	0.201	0	0	0	0	0	0	0.572	0
Khagrachori	0.088	0	0	0.128	0.006	0	2.282	0.279	0.510
Rangamati	0.019	0	0	0.076	0	0	0.643	0.387	0.050
Bandorban	0.008	0	0	0.067	0.010	0	0.985	0.307	0.024
Jessore	0.010	0	0	0.094	0.008	0	0.846	0.183	0.010
Savar	0.005	0	0	0.095	0.013	0	1.462	0.316	0.024

Table 5: Transfer Factor (TF) of heavy metals in rice samples grown in different districts.

The observed arsenic levels in rice samples from all sample collection areas were lower than world safe limit of 5 mg/kg (FAO/WHO, 2004). Statistical analysis for arsenic level showed that there was a significantly ($P < 0.001$) highest level of arsenic (0.68 mg/kg) in rice sample of Naogaon area. Arsenic contents in rice from Savar, Rangamati and Bandorban were not significantly different. Lowest level of arsenic was found in Satkhira rice. No arsenic was detected in rice from Joypurhat area. Meharg & Rahman (2003) found high concentration of arsenic level (1.836 mg/kg) in three grain samples collected from Naogoan district of Bangladesh. In a study Shaban et. al. (2012) reported As concentration in Aman rice of Sylhetdistrict of Bangladesh was 85.7 µg/kg, which is lower than the present study. On the other hand the As concentration reported by Williams et al. (2009) in rice from the regions (contaminated area) ranged from 0.26to 0.42 mg/kg. Rice samples from only three areas (Joypurhat, Naogaon and Shatkhira) showed lead content and among them, Shatkhira showed significantly ($P < 0.001$) lowest (1.35 mg/kg) lead content compare to other two areas which was lower than world safe limit of 2 mg/kg (WHO 1985). Among all eight experimental areas, Shatkhira and Jessor showed significantly ($P < 0.001$) highest (60.77 mg/kg) and lowest (16.82 mg/kg) zinc content in rice sample respectively. All observed values of zinc were lower than world standard limit of 150 mg/kg (FAO/WHO, 1984). In a study, Williams *et al.* (2009) reported Zn concentration in rice of Gazipur, Jessore and Faridpur were 16.85, 13.01 and 14.98 mg/kg respectively. Rice from khagrachori and Rangamati showed highest value (32.55 and 32.98 mg/kg respectively) of manganese with no significant difference between the locations. A significantly lowest level of manganese content was found in rice sample from Jessore area. Mn contents in all samples from eight experimental locations were found to be higher than world standard limit of 5.4 (FAO/WHO, 1984). The concentration of Mn in the present study agrees with the value found by Shaban *et al.* (2012). Maximum levels of cadmium were found in rice from Khagrachori (2.83 mg/kg) and Savar (2.89 mg/kg) areas than other areas and the observed limits were higher than world safe limit of 1 mg/kg (FAO/WHO, 1984). In a study Shakerian *et al.* (2012) found concentration of Pb and Cd in rice grain in central Iran is 0.062 and 0.068 mg/kg which is much lower than the rice of Bangladesh. There was no significant difference in cadmium level between Bandarban and Jessor rice. ANOVA analysis showed that a lowest level of cadmium (0.06 mg/kg) was found in rice from Joypurhat and Naogoan. In case of Co the concentration was too low to determine with an exception of rice sample of Shatkhira.

3.4   Spatial Distribution of Heavy Metal in Rice

The Spatial distribution of heavy metals Cr, Co, Ni, Cu, As, Pb, Cd, Zn and Mn are shown in the Figure 2. Cr was distributed more or less all the areas studied but the highest level was found for Khagrachori in North-East zone. Highest Co level was found in the South zone and the highest Ni level was found in the North zone. No regular distribution for Cu, As, Zn and Mn in rice was found. Huang *et al.* also reported that spatial patterns of heavy metals in rice were irregular in their geographical distribution (Huang *et al.*, 2009). In case of Cd the highest level was found in the hill tract area of North-East zone, whereas the highestPb level was observed in the North zone of Bangladesh.

3.5   Transfer Factor of Metals from Soil to Rice

Table 5 summarizes the metal transfer factor (TF) in rice samples from different districts. The TF value for Cr, Co, Ni, Cu, As, Pb, Cd, Zn and Mn for rice samples varied greatly between different districts. The difference in TFs between

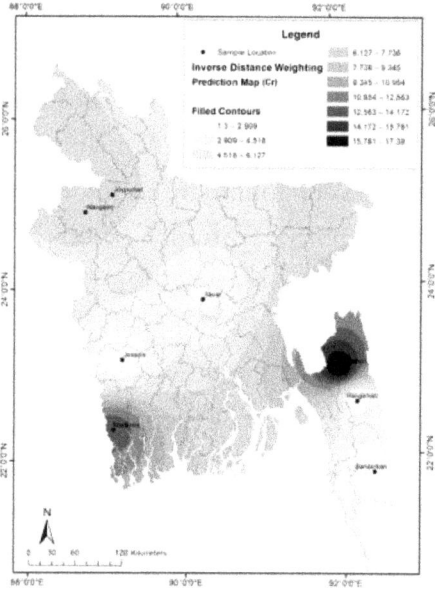

Cr

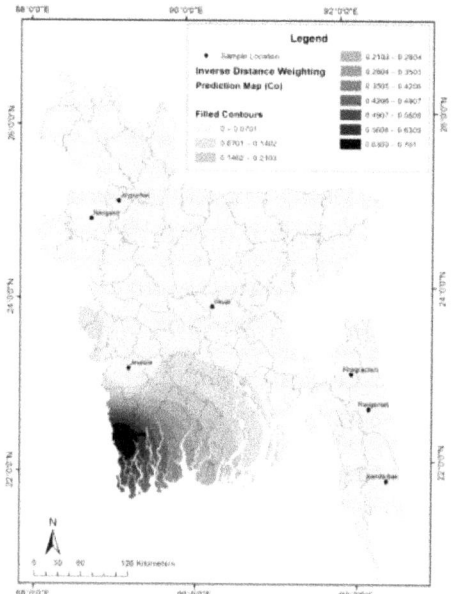

Co

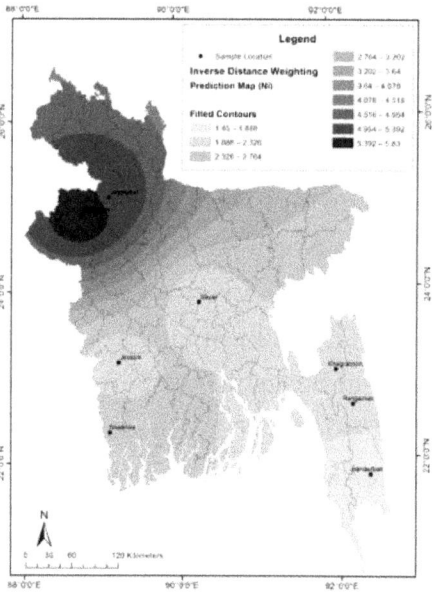

Ni

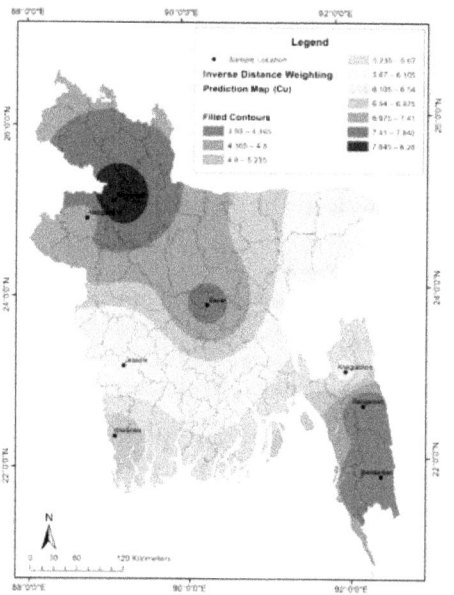

Cu

Continued on next page (Figure 2)…

... Continued from previous page (Figure 2)

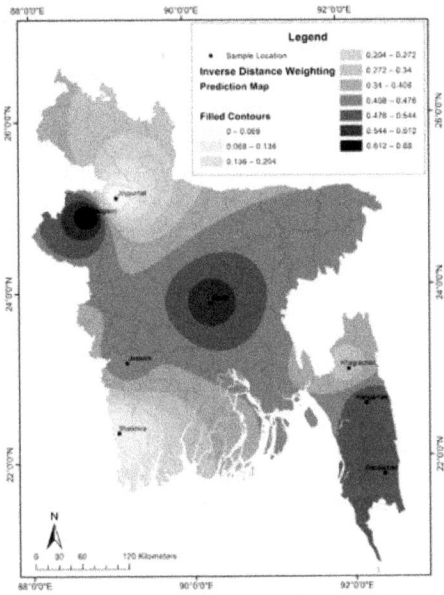

As

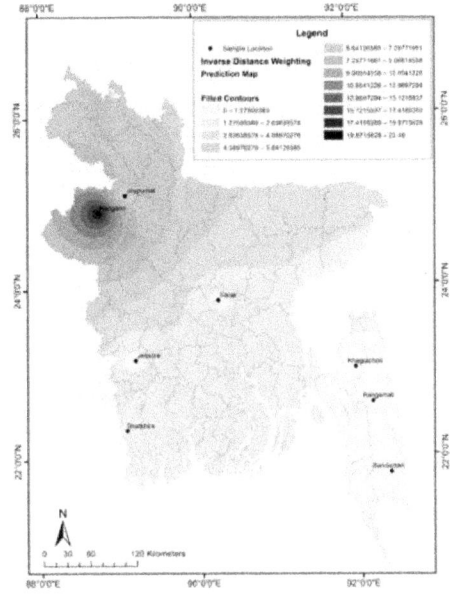

Pb

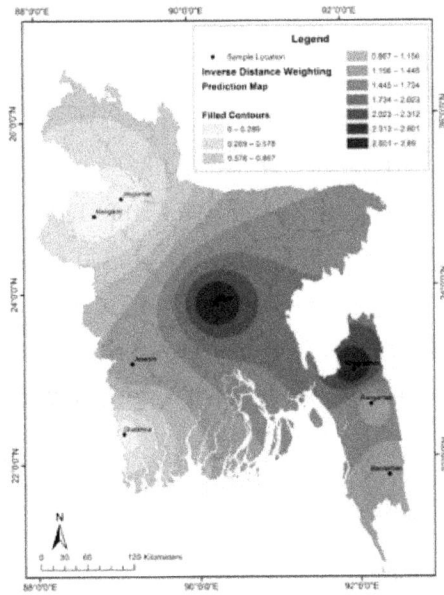

Cd

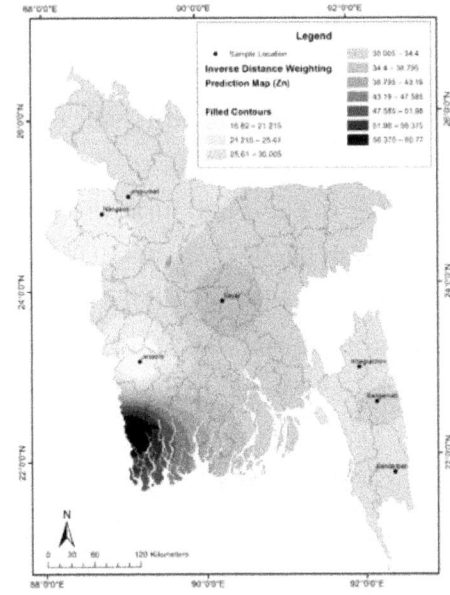

Zn

Continued on next page (Figure 2)...

... Continued from previous page (Figure 2)

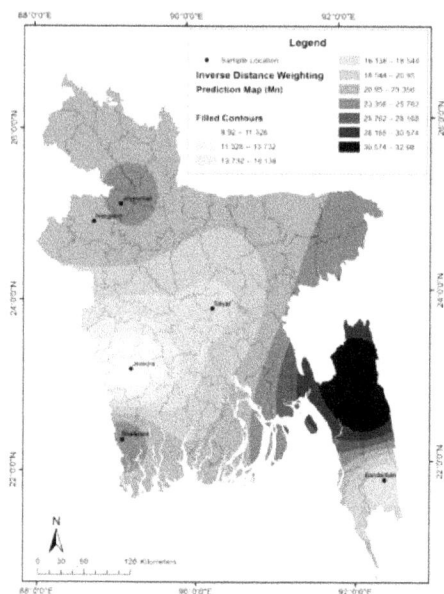

Mn

Figure 2: The spatial distribution of Cr, Co, Ni, Cu, As, Pb, Cd, Zn and Mn in rice from different district of Bangladesh.

locations may be related to soil nutrients management and soil properties. The TF value for Cr varied from 0.005 to 0.201. The Highest TF value of Cr was 0.201 found for rice in Shatkhira district. For Cu, As, Cd, Zn and Mn, the TF value ranges from 0.67 – 0.216, 0 – 0.018, 0 – 2.282, 0.183 – 0.572 and 0 – 0.510 respectively. On the other hand TF value for Co and Ni was calculated 0 (zero) because in most cases Ni and Co concentration in rice was too low to detect. For all the heavy metal analysed TF value showed a lower value with an exception of Cd. TF value for Cd followed a trend as Khagrachri > Savar > Bandorban > Jessore>Rangamati > Jopurhut = Noagaon=Shatkhira. TF value for lead showed zero value for all the samples except Joypurhut rice. Hence the zero TFvalue indicates that source of those elements are not soil.

3.5   Metal Pollution Index(MPI)

Metal Pollution Index (MPI) for rice as shown in Figure 3 followed a decreasing sequence of JH > KC > NG > SV > RM > SK > BB > JS.Metal pollution index (MPI)

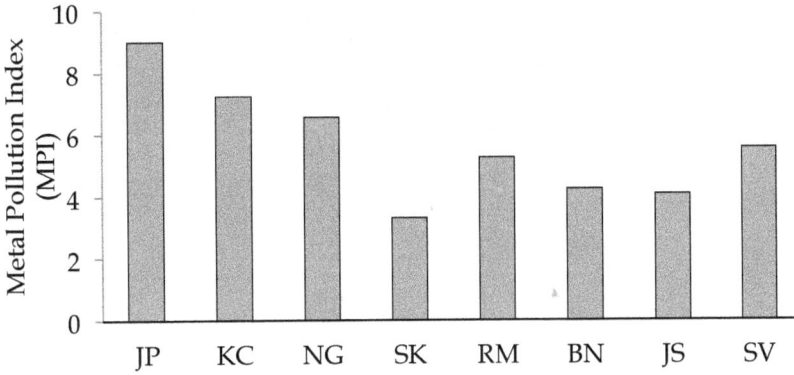

| JP: Joypurhat | KC: Khagrachori | NG: Noagaon | SK: Shatkhira |
| RM: Rangamati | BN: Bandorban | JS: Jessore | SV: Savar |

Figure 3: Metal Pollution Index (MPI) in rice samples.

is suggested to be a reliable and precise method for metal pollution monitoring of wastewater irrigated areas(Usero *et al.*, 1997). However higher MPI of rice from Joypurhat, Khagrachori, Noagaon, Savar and Rangamati suggested that these may cause more human health risk due to higher accumulation of heavy metal in rice.

3.6   Daily Intake of Metal(DIM) by the Consumption of Rice

Values for DIM calculated for adult (70 kg)are presented in Table 6. Daily intake of metal for Cr, Co, Ni, Cu, As, Pb, Cd, Zn and Mn ranged from 0.0081 – 0.1088, 0 – 0.0044, 0.0091 – 0.0365, 0.0246 – 0.0518, 0 – 0.0043, 0 – 0.0157, 0 – 0.0074, 0.1052 – 0.3802 and 0.0558 – 0.2064 mg/person/day respectively. These data revealed that the values of daily intake of Cr, Mn, Cu and Zn are within the recommended value suggested by different organization (Farid *et al.*, 2004; Ogabiela *et al.*, 2011; Dawd, 2010). For As, Pb and Cd,DIM was found higher than the permissible value.In a study Zhu Huang *et al.*, (2013) reported their daily estimated mean take of As, Pb and Cd were 1.02, 1.26 and 0.64 µg/kg for adult from Zhejiang province of China, which are lower that the value presented in this paper.However the higher DIM value for As, Pb, and Cd is an indication of high exposure of those elements for long term large consumption of rice.

Sample	DIM mg/person/day								
ID	Cr	Co	Ni	Cu	As	Pb	Cd	Zn	Mn
Joypurhat	0.0389	0	0.0335	0.0518	0	0.0100	0	0.1719	0.1561
Naogaon	0.0399	0	0.0365	0.0457	0.0043	0.0157	0	0.1340	0.1310
Shatkhira	0.0850	0.0044	0.0155	0.0312	0.0004	0.0084	0	0.3802	0.1556
Khagrachori	0.1088	0	0.0122	0.0361	0.0019	0	0.0177	0.1703	0.2037
Rangamati	0.0129	0	0.0126	0.0246	0.0034	0	0.0074	0.1933	0.2064
Bandarban	0.0081	0	0.0091	0.0251	0.0034	0	0.0083	0.1726	0.1063
Jessore	0.0097	0	0.0128	0.0366	0.0028	0	0.0086	0.1052	0.0558
Savar	0.0167	0	0.0121	0.0474	0.0036	0	0.0181	0.2316	0.1116
Recommended value [abcd]	0.05 – 0.20	–	–	2.00 – 3.00	0.0014	0.0012	0.0001	12.00 – 15.00	2.00 – 5.00

Table 6: Daily intake of metal by consumption of rice samples. a = Farid, *et.al.* (2004); b = Ogabiela *et.al.* (2011); c = Dawd, G. (2010);d= Huang *et.al.* (2009)

3.7   HRI of Heavy Metals

The health risk index for heavy metals by consumption of rice of different district of Bangladesh for adult was calculated and values are given in the Table 7.

The result revealed that health risk index (HRI) for Cr (0.005 – 0.073) and Co (0 – 0.001) for rice grown in the district concern are lower than 1 indicating no risk. HRI of Ni, Cu, Zn ranging from 0.454 – 1.824, 0.615 – 1.295 and 0.350 – 1.267, which are acceptable but a higher HRI value was observed for As (0 – 014.183), Pb(0 – 4.442) and Cd (0 – 18.083) in most of the region.Silva.*et al*, (2005) reported that the daily intake of Cd was estimated 25 – 60 μg for a 70 kg person from uncontaminated areas. The highest HRI for As, Cd and Pb was associated with the rice sample for Noagaon, Joypurhut and Savar district respectively. In case of Mn, the HRI value(1.691 – 6.253) was quite high but hence not pose any potential risk as Mn is an essential element for human health.

Sample ID	Health Risk Index								
	Cr	Co	Ni	Cu	As	Pb	Cd	Zn	Mn
Joypurhat	0.026	0	1.674	1.295	0	4.442	0	0.573	4.729
Naogaon	0.027	0	1.824	1.142	14.183	3.895	0	0.447	3.969
Shatkhira	0.057	0.001	0.776	0.781	1.251	2.112	0	1.267	4.716
Khagrachori	0.073	0	0.610	0.903	6.257	0	17.707	0.568	6.172
Rangamati	0.008	0	0.632	0.615	11.263	0	7.446	0.644	6.253
Bandarban	0.005	0	0.454	0.627	11.263	0	8.259	0.575	3.221
Jessore	0.006	0	0.641	0.917	9.177	0	8.572	0.350	1.691
Savar	0.011	0	0.607	1.186	12.097	0	18.083	0.772	3.382

Table 7: Health Risk Index (HRI) for heavy metals in Rice of different district.

3.8 Hazard Index (HI)

The Hazard Index (HI) has been calculated for the toxic element Cr, Co, As, Pb and Cd in rice samples of different district. The calculated value for HI followed the order of Savar > Khagrachori > Bandorban > Rangamati > Noagaon > Jessore > Joypurhat > Shatkhira. When the Hazard index exceed 1.0 there is concern for potential health effect. Even though there was no apparent risk when each metal was analysed individually, the potential risk could be multiplied when considering all heavy metals. From overall assessment it can be revealed that rice from Shatkhira and Joypurhat district is more safe for cosume, whereas, rice from Noagoan, Khagrachori, Rangamati, Bandorban, Jessore and Savar are very hazardous.

4 Conclusion

The extend of heavy metal contamination varied with the metal species and plant types. The toxicity of the metal in agricultural product depends on the relative level of exposure of crops to the contaminated soils as well as the deposition of toxic elements in the polluted air by sedimentation. However in the present study, the level of Co, Ni, Cu, Pb, Zn, and Mn in soil samples analyzed showed a lower value than the World Average Value (WAV)and European Union Standard (2006) except As and Cr. On the other hand Cd showed a value lower than the standard set by European Union Standard but higher than World average value. The levels of heavy metal in the water collected from the rice field were within the permissible level of irrigation water suggested by department of Environment Dhaka, Bangladesh. During the study it was found that concentration of most of the elements in rice samples are quite high compared to FAO/WHO permissible limit except Cu. Cao. H, *et al.* reported that the concentration of Cr, Cu, Pb, Cd, and Zn in the rice of an industrial zone in Jiangsu, China was 0.75, 2.64, 0.054, 0.014, and 12.0 mg kg^{-1}, which are much lower than the Bangladeshi rice. Data presented in table 4 demonstrated that Cr, As, Pb and Cd can pose substantial risk to the consumer. Hence Cr has the biggest non-carcinogenic effect while Pb, Cd, As generates the greatest cancer risk. The calculated Health Risk Index (Table 7) indicate that there is no risk for Cr but in most cases HRI value for As, Cd and Pb is many fold higher than the safe value.

Soil to plant transfer is one of the key components of human exposure to metal through food chain. The present data revealed that TF values differed significantly between locations. The difference in TF between locations may be related to soil nutrient management and soil properties. According to the TF calculated in this study it can be concluded that Rice plants are high Cd-accumulators as in most cases TF value for Cd are quite high compared to other elements. In a study Reeves and Chaney (2001), reported that Cd, a toxic element is of primary concern in soil and food contamination, particularly in rice cropping system. Hence , basic staple food such as rice and wheat can accumulate relatively high amounts of Cd when grown in contaminated soil (Silva. *et al.*, 2005).The dietary Cd absorption rate in human has been estimated at 5% of its total intake (WHO, 1989; IPCS, 1992). Satarug *et al.* (2004) reported that acuate toxicity caused by Cd contaminated food is very unusual but chronic exposure may be frequent. However, from overall study it can be concluded that the toxic metal accumulation in rice was severe and has potential to cause health risk to the consumer. It is suggested to regular monitoring should be enforced as these metal accumulation in rice.

Acknowledgements

The authors appreciated the help and association of the stuff members of Chemistry Division, Atomic Energy Centre, Dhaka. Special thanks to IAEA for their support regarding this work.

Conflict of Interests

The authors declare that there are no conflict of interest regarding the publication of this paper.

Authors' contribution

Sample was collected by M.K. Munshi, M. Islam, A. Khatun and A. Hossain, analysis was carried out by Y.N. Jolly, R. Huque, A. Islam, M.S. Rahman, S. Akter, J. Kabir. Final manuscript was drafted and edited by Y.N. Jolly and R.Huque. All authors read and approved the final manuscript.

References

Bangladesh among the top rice-eaters, The daily Star, Bangladesh Monday, December, 12, 2011.

Chaney, R.I. (1980). Health risk associated with toxic metals in municipal sludge in: Bilton G et al (ed) Sludge –Health-Risk of land application. Ann Arbor Scie. Publi MI, pp. 59-83.

Cui, Y.J., Zhu, Y.G., Zhai, R.H., Chen, D.Y., Huang, Y.Z., Qui, Y. & Liang, J.Z. (2004). Transfer of metals from soil to vegetables in an area near a smelter in Nanning, China. Environment International, 30, 785-791.

Dawd, G. (2010). Determination of the levels of essential and toxic metal constituents in cow's whole milk from selected sub-cities in Addis Ababa, Ethiopia, M. Sc, Thesis, Addis Ababa University, Ethiopia.

Environmen Agency (2009). Using Soil GuidelineValues, Science Report SC050021/SGV Introduction. Bristol: Environment Agency.

Fu, J., Zhou, Q., Liu, J., Liu, W. & Wang, T. (2008). High levels of heavy metals in rice (Oryza sativa L.) from a typical E-waste recycling area in southeast China and its potential risk to human health. Chemosphere, 71, 1269–1275.

Friberg, L., Nordberg, G.F. & Vpuk, B. (1994). Handbook on the Toxicity of Metals. Elsevier. North Holland, Bio Medical Press, Amsterdam.

Food and Nutritional Board (2004). Dietary Reference Intakes [DRIs]. Recommended Intake for Individuals. National Academy of Sciences. Washington, DC: USA.

Fertilizer Recommendation Guide (2005). Soil fertility status of different agro ecological zone.Bangladesh Agricultural Research Corporation. http://www.dae.gov.bd/pdf/soil-fertility-status.pdf. Access 04/01/2014

FAO/WHO (1990). Food standards programme, Guideline levels for cadmium and lead in food. Codex committee of food additives and contamination, 22nd session, Haugue, the Netherland.

FAO/WHO (1984). List of maximum levels recommended for contaminants by the Joint FAO/WHO Codex Alimentarius Commission. Second Series. CAC/FAL, Rome, 3, 1–8

Farid, S.M., Enani, M.A. & Wajid, S.A. (2004). Determination of trace elements in cow's milk in Saudi Arabia., JKAU: Eng Sci., 15(2), 131-140.

Huang, X., Wang, H., Zhou, J., Ma, C. & Du, C. (2009). Risk assessment of potentially toxic element pollution in soils and rice in a typical area of the Yangtze River Delta. Environ Pollut., 157, 2542-2549.

IPCS-International Programme on Chemical Safety (1992). Cadmium. Environmental health Crieteria 134, Geneva. World Health Organisation.

Industrial Effluent Quality Criteria (1999). A comparison of Environmental Laws, Bangladesh Gazette Additional 28, Department of Environment (DoE), the Ministry of Environment and Forests, Bangladesh, pp 60.

Jolly, Y.N., Islam, A. & Akbar, A (2013a). Transfer of metals from soil to vegetables and possible health risk assessment, SpingerPlus, 2, 385. doi; 10,1186/2193-1801-2-385.

Jolly, Y.N., Akter, S., Kabir, J. & Islam, A. (2013b). Health risk assessment of heavy metals via dietary intake of vegetables collected from an area selected for introducing a Nuclear Power Plant. Res. J. Phy. and App. Sci., 2(4), 43-51.

Meharg, A.A. & Rahman, M.M. (2003). Arsenic contamination ofBangladesh paddy field soils: Implications for rice contribution to arsenic consumption. Environ. Sci. Technol., 37, 229-234

Ogabiela, E.E., Udiba, U.U., Adesina, O.B., Hammuel, C., Ade-Ajayi, F.A., Yebpella, G.G., Mmereole, U.J. & Abdullahi, M. (2011). *Assessment of metal levels in fresh milk from cows grazed around Challawa Industrial Estate of Kano, Nigeria. J. Basic Appl. Sci. Res., 1(7), 533-538.*

Pendias, A.K. & Pendias, H. (2000). *Trace elements in soils and plants, 3rd Ed. CRC press, LLC, 4, 10-11.*

Reeves, P.G. & Chaney, R.L. (2001). *Minereal nutrient status of female rats affects the absorption and organ distribution of cadmium from sunflower kernels (Helianthusannuus L.,) . Environmental Research, 85, 215-225.*

Rahman, M.S., Molla, A.H., Saha, N., Rahman, A. (2012). *Study on heavy metals levels and its risk assessment in some edible fishes from Bangshi River, Savar, Dhaka, Bangladesh, Food Chemistry, 134, 1847-1854.*

Rahman, M.S., Saha, N., Molla, A.H. (2014). *Potential ecological risk assessment of heavy metal contamination in sediment and water body around Dhaka export processing zone, Bangladesh. Environ Earth Sci., 71(5), 2293-2308.*

Shaban, W., Al-Rmalli, Jenkins, R.O., Watts, M.J. & Haris, P.I. (2012). *Reducing human exposure to arsenic and simultaneously increasing selenium ans zinc intakw by substituting non-aromatic rice with aromatic rice in the diet, biomedical Spectroscopy and Imaging, 1, 365-381.*

Smith, C.J., Hopmans, P. & Cook, F.J. (1996). *Accumulation of Cr, Pb, Cu, Ni, Zn and Cd in soil following irrigation with untreated urban effluents in Aust. Environ pollut., 94(3), 317-323.*

Saleemi, M.A. (1990). *In: Saleem, Y. (Ed). Environmental Scenario of Lahore; In brief of Selected Articles and Lectures. EPA, Punjab, Lahore, 44-48.*

Silva, A.L., Barrocos, O.D., Jacob, P.R.G. & Moreira, J.C. (2005). *Dietary intake and health effects of selected toxic elements. J. Plant Physiol., 17(1), 79-93.*

Satarug, S., Moore, M.R. (2004). *Adverse health effects of chronic exposure to low-level cadmium in foodstuffs and cigarette smoke, Environ. Health Perspect, 112, 1099-1103.*

Ureso, J., Gonzalez-Regalado, E. & Gracia, I. (1997). *Trace elements in bivalvemollusks Ruditapes decussates and Ruditapesphillippinarum from Atlantic Coast of Southern Spain. Environ. Int., 23(3), 291-298.*

Shakeria, A., Rahimi, E. & Ahmadi, M. (2012). *Cadmium and lead content in several brands of rice grains (oryza sativa) in central Iran. Toxicol. Ind. Health, 28, 955-969.*

US. Environmental Protection Agency (US EPA) (1997). *Exposure Factors Handbook.*

General Factors, EPA/600/P-95/002Fa, vol.1.Office of Research and Development, National Centre for Environmental Assessment. US Environmental Protection Agency, Washington, DC, http://www.epa.gov/ncea/pdfs/efh/front.pdf

U.S. Environmental Protection Agency (US EPA) (1989). Risk Assessment Guidance for Superfund: Human Health Evaluation Manual [part A]: Interim Final.U.S. Environmental Protection agency, Washington, DC, USA [EPA/540/1-89/002.

U.S. Environmental Protection Agency (1998). Integrated Risk Information System (IRIS) on Arsenic. National Center for Environmental Assessment, Office of Research and Development, Washington, DC.

U.S. Environmental Protection Agency (US EPA) (2002). Region 9, Preliminary Remidation Goals. http://www.epa.Gov/region09/waste/sfund/prg

US-EPA IRIS (2006). United States, Environmental Protection Agency, Integrated Risk Information System. http://www.epa.gov/iris/substS

Wagner, G.J. (1993). Accumulation of cadmium in crop plants and its consequences to human health. Adv. Agron, 51, 173-212.

Williams, P.N., Islam, S., Islam, R., Jahiruddin, M., Adomako, E., Soliaman, A.R.M., Rahman, G.K.M.M., Lu, Y., Deacon, C., Zhu, Y. & Meharg, A.A. (2009), Arsenic limits trace mineral nutrition (selenium, zinc and nickel) in Banglades rice grain, Environ. Sci. Technol., 43, 8430-8436.

World Health Organization [WHO] (1993). Evaluation of Certain Food Additives and Contaminants. In: Forty-First Report of the joint FAO/WHO Expert Committee on Food Additives, WHO. Geneva, Switzerland, (WHO Technical Series, 837).

WHO (1985). Guidelines for DrinkingWater Quality. Recommendation, WHO,Geneva, Switzerland.

WHO (1996). Permissible limits of heavy metals in soil and plants, (Genava: World Health Organization), Switzerland.

World Health Organisation (WHO)(1989). Evaluation of certain food additives and contaminants (Thirty-third report of the joint FAO/WHO expert committee on food additives). WHO Technical Report Series No. 776. World Health Organisation, Geneva.

Zhao, K., Liu, X., Xu, J., Selim, H.M. (2010). Heavy metal contaminations in a soil–rice system: Identification of spatial dependence in relation to soil properties of paddy fields. J. Hazard. Material., 181, 778-78.

Impact of Magnetic Field on Crop Plants

K.N.Guruprasad[1], M.B.Shine[1] and Juhie Joshi[1]

1 Introduction

The Earth's magnetic field is a natural component of the environment and an inescapable environmental factor for living organisms including plants. The intensity of geomagnetic field is about 50 μT (micro Tesla) (35 μT near the equator, 70 μT near the magnetic poles of the Earth (Zhadin, 2001)) in which plant and seedling growth generally takes place. If the magnetic environment is changed the development of plant may be altered. The influence of the geomagnetic field on the growth of plants was scientifically established for the first time in 1862 by the French chemist Louis Pasteur (1822 – 1985), during his experiments on fermentation, when he discovered that the Earth's magnetic field had a stimulating effect on that process. Now a day's the role of magnetic fields and their influences on functioning of biological organisms are being actively studied and this discipline is called magnetobiology.

Magnetobiology is a new synthetic discipline encompassing the principles and techniques of many sciences, from engineering, physics, chemistry, biology and centered around biophysics. It is an approach in radiobiology of non-ionizing radiation; the line of investigation in biophysics that studies biological effects of oscillating or static and low-frequency magnetic fields, which do not cause heating of tissues. Magnetobiology has made notable advances only in the last 10 – 20 years in plant science. Now there is growing evidence that magnetic treatment of seeds before sowing, allows spending less on the seed as germination rates are increased substantially in the same seed lots. Some of the studies with seedlings of different plant species placed in the magnetic field show that

[1] School of Life Sciences, Devi Ahilya University, India

their growth enhanced (Vashisth & Nagarajan, 2010; Shine *et al.*, 2011a; 2011b; Shine & Guruprasad, 2012) while others show that development is inhibited (Huang & Wang, 2007). Hence, it may be predicted that seeds and plant react differently at different frequencies and different intensity of magnetic fields. Magnetic fields have effect on plant and seeds based on the field intensity, exposure time, signal form, flux density and source frequencies.

Various agronomic practices have often been employed to improve the performance of crop plants in terms of growth and yield. Physical treatment of seeds under low magnetic fields is one of the successful methods which has shown to improve the growth and yield of plants like maize (Shine & Guruprasad, 2012), rice (Florez *et al.*, 2004), mungbean (Huang & Wang, 2008) and sunflower (Vashisth & Nagarajan, 2010). In these plants pretreatment of seeds with magnetic field enhances germination percentage, seedling vigor, vegetative growth and yield of plants.

Even though the interactions are not fully understood, electromagnetic fields have been known to act as bio-stimulators for the growth of many types of plants and seeds. The magnetic field may provide a feasible non-chemical solution in agriculture. At the same time it offers advantages to protect environment and also safety for the applicator. So, in the present review we have discussed effects of magnetic field on crop plants.

2   Germination and Seedling Growth

Crop yields can be maximized by establishment of an adequate and uniform plant population for which good quality seed is a pre requisite. The gains from agronomic inputs are drastically reduced if the seed is of poor quality resulting in a poor stand. The application of high quality sowing materials which has been properly pre-prepared is an important yield enhancing factor in plant cultivation. Most often chemical methods consisting seed dressing, priming with various chemical substances are used in the pre-sowing seed treatment. Such methods are considered as very effective in vigor improvement but are not eco-friendly and have handling problems. Interest has therefore shifted to physical treatments like gamma rays, laser, electron beam, microwave, magnetic field and radiofrequency energies to bring about bio-stimulation of seeds, which leads to increased vigor and contributes to improved development of the plants. Physical methods moreover provide significant yield improvement without the hazard of toxic fertilizer and management cost. Therefore the practical applicability of physical seed treatment for enhancing the seed performance should be standardized for commercial use. Magnetic seed treatment is one of the physical pre-sowing seed treatments especially worth attention since its impact on the seeds can change the processes taking place in the seed and stimulate plant development.

This technique has numerous practical applications in modern agriculture:

1. Enhanced seedling growth, germination rate and yield.

2. Decreases the seed rate per hectare by increasing the germination percentage, thus reduces the cost of cultivation.

3. Environment friendly: no adverse effect on the environment.

4. Alleviates the deteriorating effect of storage and help in salvaging costly seed

5. Helps in establishing a uniform stand as a result of improved vigor and less seed rate and thus improves crop yields.

Gubbels (1982) observed that a small number of seed lots produced earlier and more vigorous seedling growth as a result of magnetic treatment, but the differences were small and inconsistent. In the field, there were no improvements in yield from magnetic seed treatment in three out of four years. In one year there was a significant increase in yield for sunflower.

Exposure of maize for 2 – 10 min to magnetic field ranging from 0.06 to 0.2 T (Tesla) stimulated germination and increased the harvest by 29.5 % (Antonow *et al.*, 1982). Phirke *et al.* (1990) studied the germination of sunflower crop exposed to different magnetic field of low strength (7 – 78 gauss) and at different speed of rotation and found that the higher speed of rotation (800, 900 and 1000 rpm) enhanced seed germination percentage. Higher speed than these showed reduction in seed germination, whereas at lower speed the results were at par with the control. Positive effects of magnetic fields have been observed on wheat cultivars (Pietruszewski, 1993) and two varieties of sunflower (Kiranmai, 1994). Kiranmai (1994) studied the mutagenic effect of magnetic field in two varieties of *Helianthus annus* L. exposed to 1000, 2000 and 3000 G (gauss) for over

90 min. Amongst the three doses studied, 2000G produced positive mutations in both varieties of sunflower. Alexander & Doijode (1995) noted that aged onion and rice seeds exposed to a weak electromagnetic field for 12 h increased the germination, shoot and root length of seedlings.

Carbonell *et al.* (2000) evaluated the seeds of rice (*Oryza sativa* L.) exposed to 150 and 250 mT (milli Tesla) magnetic fields both chronically and for 20 min after seedling emergence. Chronic exposure to a 150 mT magnetic field increased both the rate and percentage of germination relative to non exposed seeds (18% at 48 h). Significant differences were also obtained for seeds exposed to 250 mT magnetic field for 20 min (12% at 48 h). Celestino *et al.* (2000) have reported accelerated sprouting rate, main shoot length, axillary shoot formation, fresh and dry weights of the emerged shoot of *Quercus suber* seedlings when exposed to chronic electromagnetic field.

Moon and Chung (2000) found that the percent germination rates, of tomato seed treated with AC (Alternating Current) electric fields ranging from 4 to 12 kV/cm and AC magnetic flux densities ranging from 3 to 1000 G exposed to 15 to 60 s time were accelerated by about 1.1 – 2.8 times compared with that of the untreated seed. However, an inhibitory effect on germination was shown in the case of the electric field more than 12 kV/cm and the exposure time more than 60 s.

Martinez *et al.* (2000) reported the influence of a stationary magnetic field on the initial stages of barley plant development. When germinating barley seeds were subjected to a magnetic field of 125 mT for different time periods (1, 10, 20, and 60 min, 24 h, and chronic exposure), increases in length and weight was observed. Maximum increases in the measured parameters were obtained when the time of exposure to magnetic field was long (24 h and chronic).

Aladjadjiyan (2002) observed that the magnetic field stimulated the shoot development of maize and led to an increase in germinating energy, germination, fresh weight and shoot length.

In broad bean and pea cultivars, the magnetic stimulation of seeds improved the sprouting and emergence of seed which resulted in higher pod number and seed yield (Podlesny *et al.*, 2004; 2005).

Growth of the germinated *Vicia faba* seedlings was enhanced by the application of power frequency magnetic fields (100 µT) that was supported by increased mitotic index and 3H-thymidine uptake (Rajendra *et al.*, 2005).

Florez *et al.* (2007) reported that exposure of maize seed to stationary magnetic field accelerated the germination and early growth of seedlings. The mean germination time and the time required for onset of germination was significantly reduced compared to control when the grass seeds: *Festuca arundinacea* Scheb and *Lolium perenne* L. (Carbonell *et al.*, 2008) and tomato seeds (Martinez *et al*, 2009) were exposed to magnetic field.

Vashisth & Nagarajan (2007; 2008; 2010) reported significant increase in speed of germination, percentage of germination, seedling vigour, in maize, chickpea and sunflower seeds exposed to static magnetic field. Magnetic field exposure of 100 mT (2 h) and 200 mT (1 h) for maize, 50 mT (2 h), 100 mT (1 h) and 150 mT (2 h) for chick pea and exposure of 50 mT (2 h) and 200 m T (2 h) for sunflower yielded the peak performance in their study with different magnetic field strength and time of exposure. The greatest increases were obtained for plants continuously exposed to 125 or 250 mT. Stationary magnetic field strengths of 50 and 200 mT for 2 h exposure increased the activities of hydrolyzing enzymes in *Helianthus anus*, which were responsible for the quick seed germination, improved seedling vigor and better root characteristics of treated seeds in this plant (Vashisth & Nagarajan, 2010).

Shine M.B. (2011 Ph.D. Thesis) showed that treatment with magnetic fields (100 mT for 120 min. and 200 mT for 60 min.) improved germination-related parameters like water uptake, speed of germination, seedling length, fresh weight, dry weight and vigour indices of maize seeds under laboratory conditions (Figure 1). Shine *et al.* (2011a) reported similar results in same parameters with magnetic fields (150 mT and 200 mT for 60 min.) on soybean seeds under laboratory conditions.

3   Free Radicals and Hydrolysing Enzymes

In seeds, reactive oxygen species (ROS) production has been considered for a long time as being very detrimental, since the works dealing with ROS were mainly focused on seed ageing or seed desiccation, two stressful situations which often lead to oxidative stress (Bailly, 2004). Numerous recent works have nevertheless brought new lines of evidence showing that the role of ROS in seeds is not as unfavorable as it was considered previously. Instead ROS may

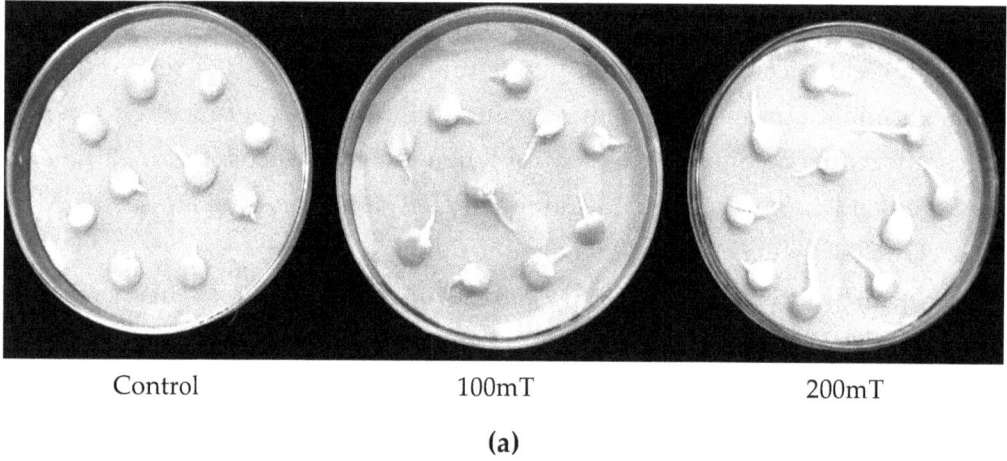

Control   100mT   200mT

(a)

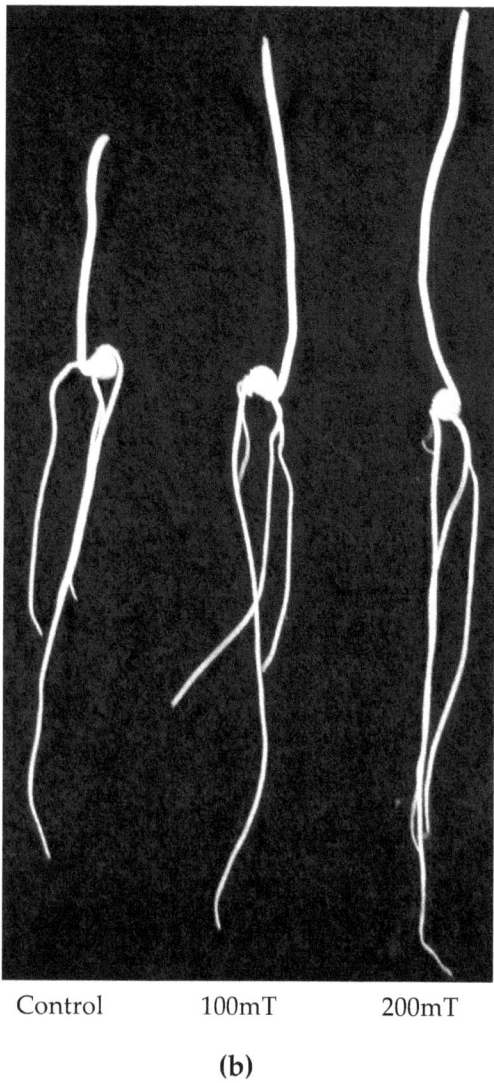

Control   100mT   200mT

(b)

Figure 1: Effect of pre-germination exposure of maize seeds to magnetic field on: **(a)** Speed of germination, and **(b)** Seedling vigour of eight day old seedlings.

play a key signaling role in the achievement of major events of seed life, such as germination or dormancy release. ROS, particularly superoxide (O_2^-) and H_2O_2 act as second messengers in many processes associated with plant growth and development (Schroeder *et al.*, 2001; Foreman *et al.*, 2003).There are few reports in soybean suggesting that seed germination is accompanied by a generation of ROS in the embryonic axis as well as in seed coat (Gidrol *et al.*, 1994; Khan *et al.*, 1996) which probably play an important role in protecting the emerging embryo against invasion of parasitic organism. Polysaccharide - splitting action of ·OH produced by peroxidase mediated reaction in the growing zone of maize seedling (Liszkay *et al.*, 2003; 2004) results in cell elongation and extension growth. ROS may play a key signaling role in the achievement of major events of seed life, such as germination or dormancy release.

Shine *et al.* (2012) reported that Magnetic field treatment enhanced the production of ROS mediated by cell wall peroxidase while ascorbic acid content, superoxide dismutase and ascorbate peroxidase activity decreased in the hypocotyl of germinating soybean seeds. An increase in the cytosolic peroxidase activity in their study indicated that this antioxidant enzyme had a vital role in scavenging the increased H_2O_2 produced in seedlings from the magnetically treated seeds. These studies contributed to the biochemical basis of enhanced germination and seedling growth in magnetically treated seeds of soybean in relation to increased production of ROS (Figure 2A, B). EPR spectrum of O_2^--PBN adduct revealed that the O_2^- radical level was lower by 16% in the leaves of plants that emerged from soybean seeds which were magnetically treated (Figure 3). Magneto-primed seeds have a long lasting stimulatory effect on plants as it reduced superoxide production in leaves and higher performance index contributed to higher efficiency of light harvesting that consequently increased biomass in plants that emerged from magneto-primed seeds (Shine *et al.*, 2011b).

Electron paramagnetic resonance spectroscopy studies in maize by Shine & Guruprasad (2012) showed that superoxide radical was reduced and hydroxyl radical was unaffected after magnetic field treatment. With decrease in free radical content, antioxidant enzymes like superoxide dismutase and peroxidase were also reduced by 43% and 23%, respectively, in leaves of plants that emerged from magnetically treated maize seeds (Figure 4).

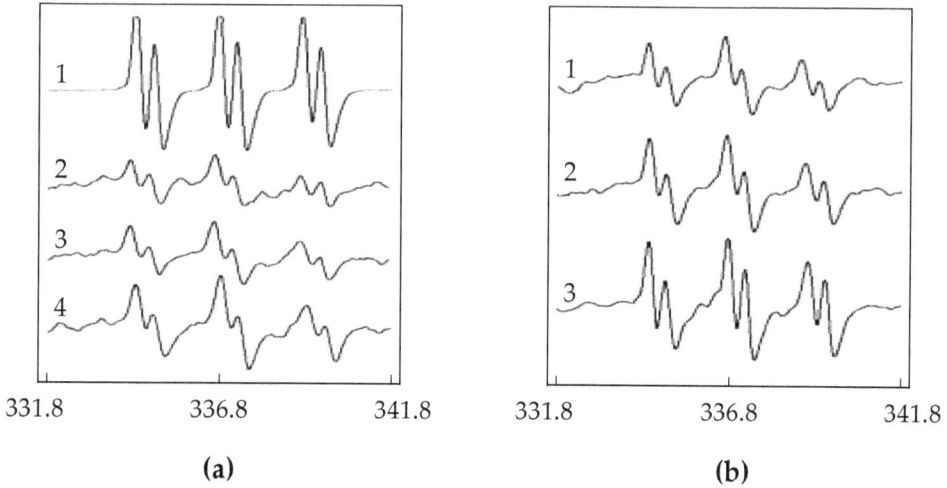

(a)   (b)

Figure 2A: EPR spectra of superoxide-PBN (phenyl N-t-butylnitrone) spin adduct from embryos and hypocotyls of soybean seedlings after magnetic field treatment. a: Embryo (1: spectra from xanthine/xanthine oxidase; 2: control; 3: 150 mT; 4: 200 mT). b: Hypocotyl (1: control; 2: 150 mT; 3: 200 mT). Each spectrum is representative of 18 individual spectra.

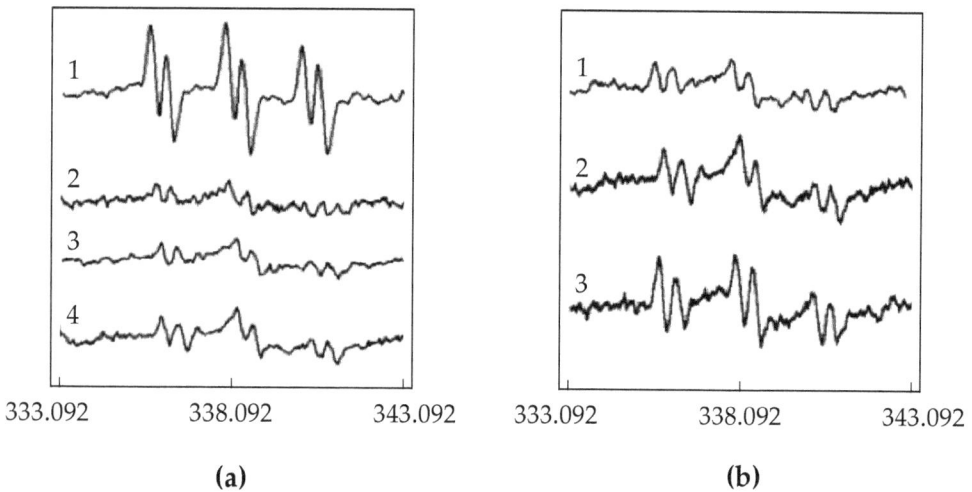

(a)   (b)

Figure 2B: EPR spectra of hydroxyethyl-POBN [a-(4-pyridyl-1-oxide)- N-tert-butylnitrone] adduct diagnostic for .OH after magnetic field treatment of soybean . a:Embryo (1: spectra from Fenton reaction; 2: control; 3: 150 mT; 4: 200 mT). b:Hypocotyl (1: control; 2:150 mT; 3:200 mT). Each spectrum is representative of 18 individual spectra.

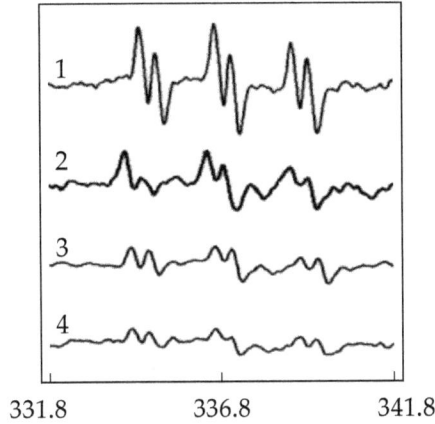

Figure 3: Effect of magneto-priming on superoxide production in soybean leaves. (1) spectra from xanthine/xanthine oxidase, (2) control,(3) 150 mT (1 h), (4) 200 mT (1 h). Each spectra is the representative of 18 individual spectra.

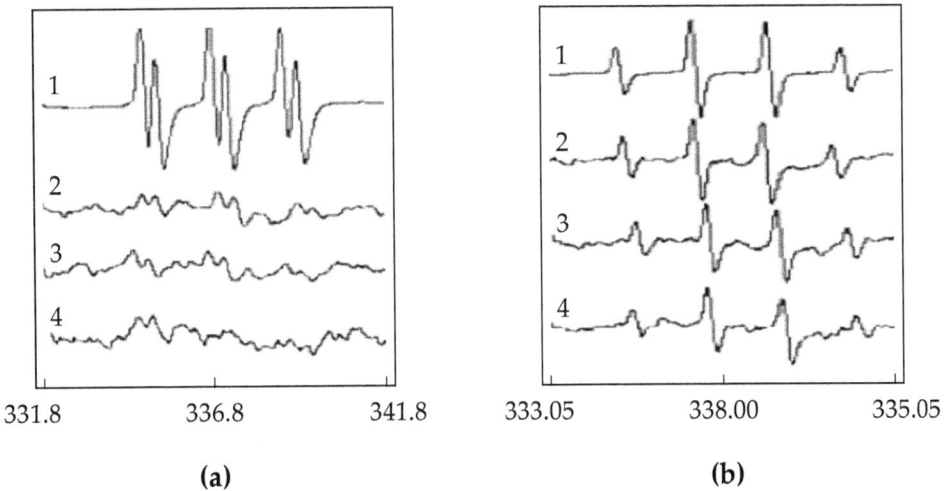

(a)   (b)

Figure 4: Effect of pre-sowing exposure of magnetic field on reactive oxygen species in maize leaves. **(a)** EPR spectra of Superoxide-PBN (Phenyl N-t-butylnitrone) spin adduct. [1 spectra from xanthine/xanthine oxidase, 2 control, 3 100 mT (2 h), 4 200 mT (1 h)]. **(b)** EPR spectra of. OH-DMPO adduct [1 spectra from Fenton reaction, 2 control, 3 100 mT (2 h), 4 200 mT (1 h)]. Each spectra is the representative of twelve individual spectra.

4   Photosynthetic Pigments and Photosynthesis

Magnetic fields increased photochemical activities in a unit of chlorophyll molecule resulting in an increase in the green pigment of wheat and bean (Lebedev *et al.*, 1977). Dhawi & Al-Khayri (2009) found that pigment content (chlorophyll *a*, chlorophyll *b*, carotenoids and total pigments) were significantly increased under static magnetic field. The highest measurements were recorded at 100 mT, after 360 min of exposure. On the other hand, alternating magnetic field decreased photosynthetic pigments content after 10 min of treatment with 1.5 T. Thus low magnetic field doses had a simulative effect on photosynthetic pigments whereas high doses had a negative effect. Chlorophyll *a* and carotenoids were more affected than chlorophyll *b*.

Shine *et al.* (2011 a) reported a positive effect on plant growth (especially leaf area, and leaf fresh weight), leaf photosynthetic efficiency, and leaf protein content of soybean apart from improved germination-related parameters. They found that Polyphasic chlorophyll *a* fluorescence (OJIP) transients from magnetically treated plants (200 mT and 150 mT for 60 min) gave a higher fluorescence yield at the J–I–P phase and total soluble protein map (SDS–polyacrylamide gel) of leaves also showed increased intensities of the bands corresponding to a larger subunit (53 KDa) and smaller subunit (14 KDa) of Rubisco in the treated plants (Figure 5A, B, C).

Similarly, growth of maize also enhanced along with up to two-fold enhancement in Performance index of the plants after magnetic field treatment. Also, phenomenological leaf model of treated plants showed more active reaction centres (Shine & Guruprasad, 2012). (Figure 6 A,B,C).

Priming of soybean seeds with static magnetic field exposure of 200 mT (1 h) and 150 mT (1 h) resulted in plants with enhanced performance index (PI) (Figure 7). The three components of PI i.e., the density of reaction centers in the chlorophyll bed (RC/ABS), exciton trapped per photon absorbed (φpo) and efficiency with which a trapped exciton can move in electron transport chain (Ψo) were found to be 17%, 27% and 16% higher, respectively in leaves from 200 mT (1 h) treated compared with untreated seeds (Shine *et al.*, 2011 b).

The ROS has a dual role in maize at two different stages of growth. At the germination and early seedling growth stage excess ROS increased in the magnetically treated seeds and enhanced speed of germination and seedling

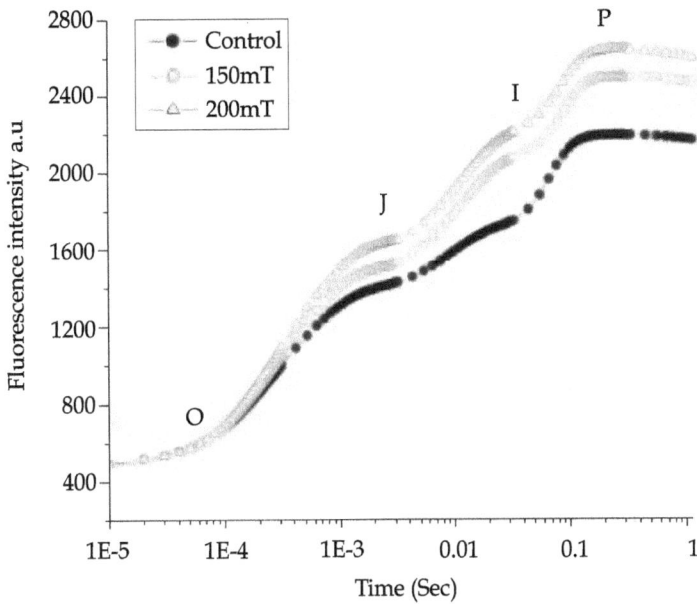

Figure 5A: Changes in polyphasic chlorophyll a fluorescence (OJIP) transient curves in soybean leaves after pre-treatment of seeds with magnetic field strengths of 150 and 200 mT for 60 min (OJIP are fluorescence yield at 20 ms, 2 ms, 30 ms, and maximum fluorescence, respectively).

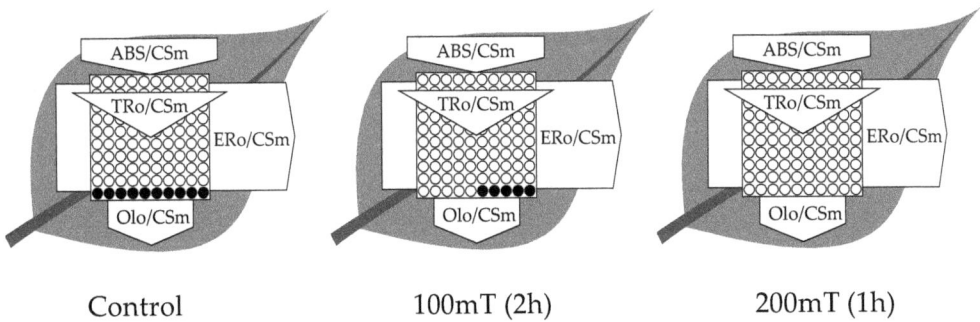

Figure 5B: Leaf model showing phenomenological energy fluxes per excited cross-section (CS) of soybean leaf. ABS/CSm ¼ absorption flux CS approximately by Fm; TR/CSM ¼ trapped energy per CS, ET/CS0 ¼ electron transport flux per CS; DI0/CSM ¼ dissipated energy per CS. Each relative value is represented by the size of the proper parameters (arrow), empty circles represent reducing QA reaction centers (active), full black circles represent non-reducing QA reaction centers (inactive or silent).

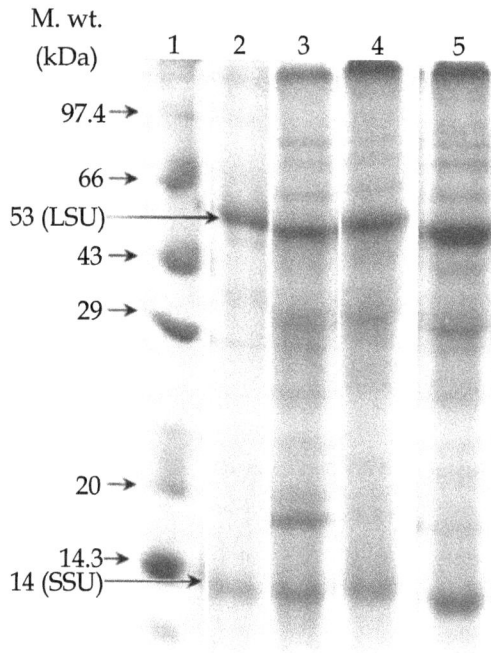

Figure 5C: Protein profiling of soybean leaf after magnetic field treatment. Lane 1: Molecular weight marker; lane 2: Partially purified Rubisco from spinach; lanes 3 – 5: Protein extracted from the leaves treated with 0,200, and150 mT (60 min), respectively.

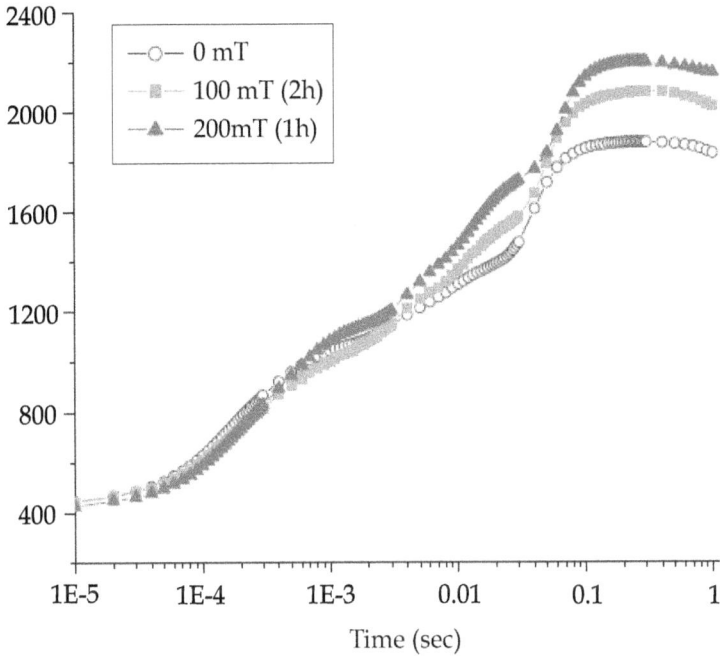

Figure 6A: Effect of pre-sowing exposure of magnetic field on Fluorescence emission transient of maize leaves, Normalized by Fo. Data are the average of fifteen individual plants from each treatment.

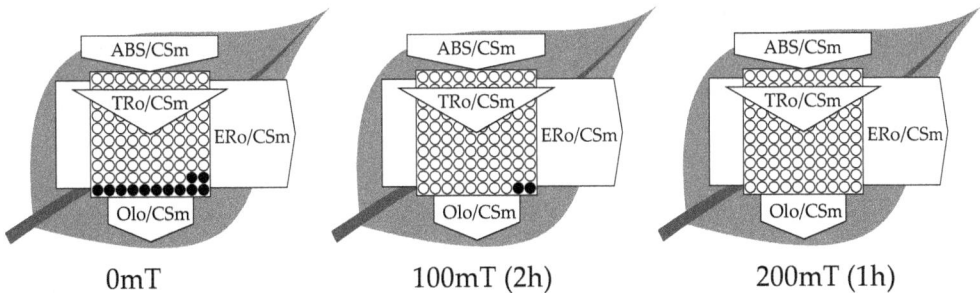

Figure 6B: Phenomenological leaf model of maize leaves showing the effect pre-sowing exposure of magnetic field. Each relative value is represented by the size of the proper parameters (arrow), empty circles represent reducing QA reaction centers (active), full black circles represent non-reducing QA reaction centre (inactive or silent).

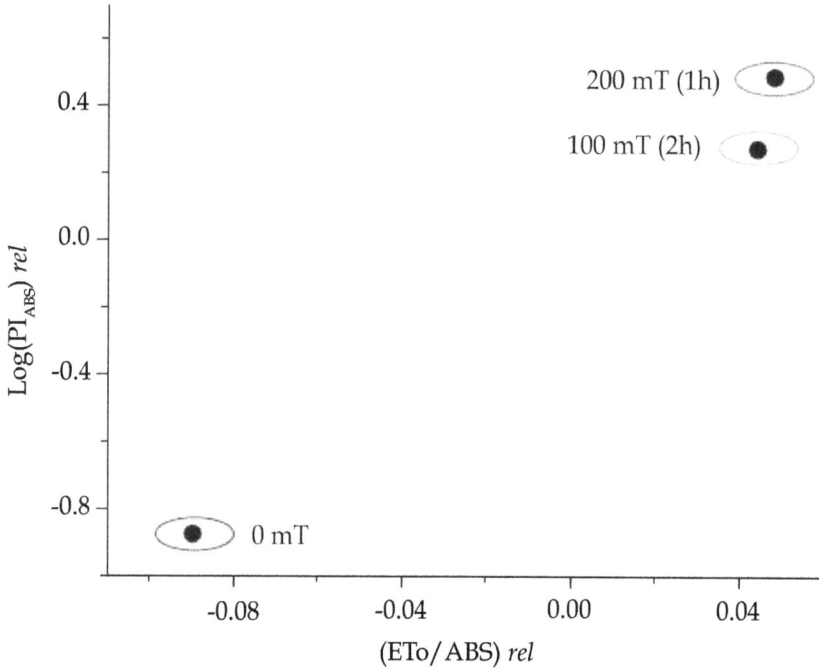

Figure 6C: Correlation of the driving force (DFABS)rel = Log (PIABS)rel as a function of the relative yield of electron transport (ET0/ABS) rel. [Log (PIABS)rel = [(PIABS)AVG treatment – (PIABS)all avg]/(PIABS)all avg. (ET0/ABS) rel = [(ET0/ABS) AVG treatment – (ET0/ABS)all avg]/(ET0/ABS)all avg]. Data are the average of fifteen individual maize plants from each treatment.

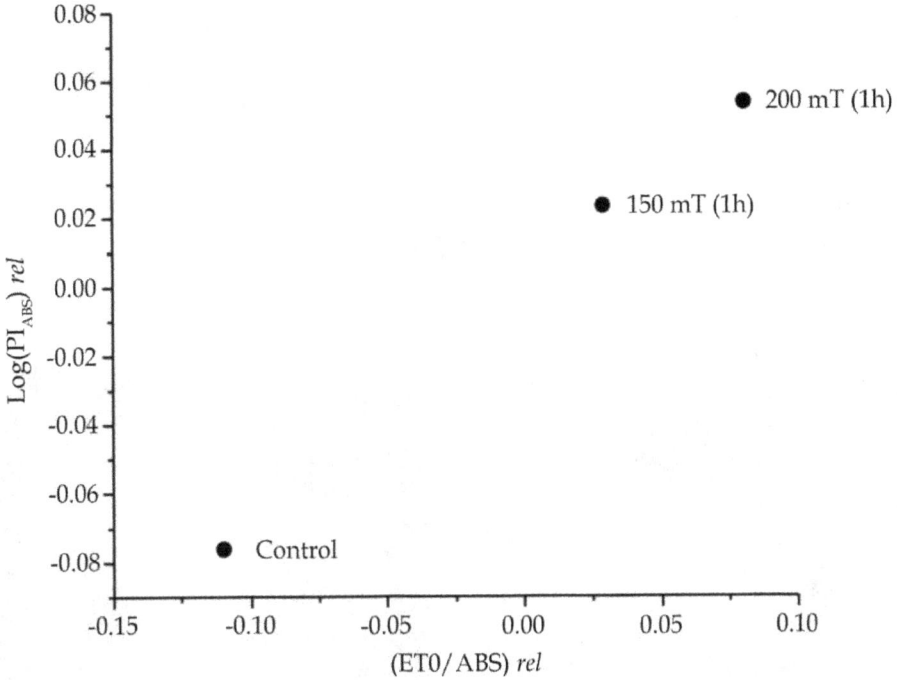

Figure 7: Correlation between the driving force (DFABS)*rel* = Log (PI-ABS)*rel* as a function of the relative yield of electron transport (ET0/ABS)*rel*. [Log (PIABS)*rel* = [(PIABS)AVG treatment - (PIABS)all avg]/ (PIABS)all avg. (ET0/ABS)*rel* = [(ET0/ABS)AVG treatment - (ET0/ABS)all avg]/(ET0/ABS)all avg in soybean.

growth. However at the later stages of growth (30 day old plants) especially in the leaves ROS content is lower after magnetic field treatment and this is accompanied by higher photosynthetic efficiency of leaves. ROS in germinating seeds could be beneficial but they are detrimental in actively photosynthetic leaves. Magnetic field treatment alters the status of ROS at both the stages of the growth to the benefits of the plants.

5   Root Growth and Nitrogen Fixation

Shabrangi *et al.,* (2010) reported positive influence on root growth of maize by magnetic field pre-sowing treatment in the first 45 days of plant growth. Vashisth & Nagarajan (2010) found that root length and root surface area showed significant increases in sunflower seedlings exposed to static magnetic fields of strength from 0 to 250 mT. Similar results have been recorded by Muraji *et al.* (1998) where in corn seedlings alternating magnetic field of 10 and 20 Hz resulted in 20% greater root growth than control plants.

Shine M.B. (2011, Ph.D. thesis) showed enhancement in root characteristics of magnetically treated maize plants. (Figure 8)

| Control | 100mT | 200mT | Control | 100mT | 200mT |

Figure 8: Effect of pre-sowing exposure of magnetic field on growth of one month old maize plants.

Plants that emerged from magnetically treated seeds of soybean showed better root characteristics (root length, fresh weight and dry weight) when compared to the untreated ones. Number and weight of root nodules per plant was also enhanced after magnetic field treatment (150 mT and 200 mT for 60 min.). The number of nodules and the fresh weight of the nodules were enhanced at 45 DAE (Days After Emergence) in soybean. Biochemical parameters viz. total protein, leghemoglobin and heme-chrome content in nodules were also enhanced by magnetic field treatment. Biochemical analysis of nodules has revealed enhancement in total soluble proteins and a specific enhancement in the level of leghemoglobin; a protein, which plays an important part in the fixation of nitrogen in the nodules (Joshi & Guruprasad, unpublished). Higher the content of leghemoglobin more will be the efficiency of the plant in terms of its capacity to fix atmospheric nitrogen (Gurumoorthi *et al.*, 2003). The data indicated a positive effect of magnetic field treatment on the activity of nitrogen fixation by enhancement in the synthesis of leghemoglobin (Joshi & Guruprasad, unpublished). Total free amino acids in leaves and nodules of soybean were reduced after magnetic field treatment. These results are in accordance with increased protein synthesis after magnetic field treatment.

Although legumes can fix nitrogen directly through the root nodules, the activity of nitrate reductase assumes importance particularly at the stage of pod filling at maturity, when the nodular activity declines. Magnetic field treatment enhanced the activity of nitrate reductase in soybean maximally at 55 DAE. This enhancement may contribute to the yield in particular. (Joshi & Guruprasad, unpublished)

Radhakrishnan & Kumari (2013) also reported enhancement in nitrate reductase in 10 day old seedlings of soybean emerged from seeds treated with pulsed magnetic field for 5 hours per day for 20 days at different frequencies. Changes in nitrate reductase activities in germinating seeds treated by electromagnets of different field strengths were also observed by Bhatnagar & Deb (1977).

6 Crop Yield

Pittman (1977) reported that pre-seeding magnetic treatment of barley (*Hordeum vulgare* L.) seed resulted in seed yield increase in 13 out of 19 field tests. Similarly treatment of spring and winter wheat seed (*Triticum aestivum* L.) resulted in yield increase in 14 out of 23 tests.

Exposure of maize for 2 – 10 min to magnetic field ranging from 0.06 to 0.2 T stimulated germination and increased the harvest by 29.5 % (Antonow *et al.*, 1982). Pietruszewski (1993) reported a positive effect of magnetic field on yield of wheat cultivars.

Phirke *et al.* (1996) reported using response surface analysis, the optimum magnetic field strength (MFS) for enhancing yield of magnetically treated seed in the field. The range for MFS was 0.072 to 0.128 T in combination with seed exposure time (SET) varying from 13 to 27 min. The study revealed that SET was found to be more effective factor than MFS for soybean, cotton and wheat seed sown in the field. An MFS of 0 .10 T was the optimal level for all three crops and SET optimal levels of 25 minutes were obtained for soybean and cotton but only 13 minutes for wheat seed. The respective yields of the three crops were increased by about 46%, 32% and 35%. It was concluded that at constant rotational speed of the disc and MFS level, SET varies from crop to crop.

Harichand *et al.* (2002) reported that field trial of 100 gauss with 40 h exposure indicated an increase in plant height, seed weight per spike and yield of wheat. An increased number of secondary branches and yield was also observed in safflower (Faquenabi *et al.*, 2009) by presowing exposure of seeds to 72 mT.

Recently, Bhardwaj J. (2012 Ph.D. thesis) reported increase in the yield of cucumber after treatment with static magnetic field. Shine M.B. (2011, Ph.D. thesis) reported increase of yield in maize after MF (magnetic field) treatment (Figure 9). Increase of yield in soybean is also observed (Joshi & Guruprasad, unpublished data).

| Control | 100mT | 200mT |

Figure 9: Representative ears of maize after magnetic field treatment.

7   Conclusions

Pre-sowing treatment of seeds under magnetic field:

- Increases speed of seed germination
- Improves seed vigour

- Improves plant growth and development

- Improves photosynthetic efficiency and increases accumulation of biomass

- Decreases free radical and hence antioxidant enzymes in leaves of treated plants

- Increases yield

8   Future perspective

Pre-sowing treatment of seeds with magnetic field has revealed beneficial effects in several crops. Exposure of seeds to magnetic field is one of the potential, safe and affordable physical pre-sowing treatments to enhance post germination plant development and crop stand. This has a potential to increase crop production per unit area of land without having any damaging effect towards any environmental component. Also, MF pretreated seeds may be effectively used for alleviation of abiotic stresses like water stress, salinity, high temperature and UV-B radiations.

References

Aladjadjiyan, A. (2002). *Study of the influence of magnetic field on some biological characteristics of Zea mays. Journal of Central European Agriculture, 3, 89–94.*

Alexander, M.P. & Doijode, S.D. (1995). *Electromagnetic field: a novel tool to increase germination and seedling vigor of conserved onion (Allium cepa L.) and rice (Oryza sativa L.) seeds with low viability. Plant Genetic Resources Newsletter, 104, 1–5.*

Antonow, G., Armjanov, N. & Todorov, T. (1982). *Untersuchungen zum Einfluß des Magnetfeldes auf die Keimenergie von Samen und den Ertrag (bulg.). Selskostopanska Techn (Sofija), 19, 5–11.*

Bailly, C. (2004). *Active oxygen species and antioxidants in seed biology. Seed Science Research, 93, 107–114.*

Bhardwaj, J. (2012). *Study of electromagnetic energies in improving vigor and storability of vegetable seeds. (Ph.D. thesis).*

Bhatnagar, D. & Deb, A.R. (1977). Some effect of pre-germination exposure of wheat seeds to magnetic fields: Effect on some physiological process. Seed Research, 5, 129–137.

Carbonell, M.V., Martínez, E. & Amaya, J.M. (2000). Stimulation of germination in rice (Oryza sativa L.) by a static magnetic field. Electro.Magnetobiol, 19, 121–128.

Carbonell, M.V., Martínez, E., Flórez, M., Maqueda, R., López-Pintor, A. & Amaya, J.M. (2008). Magnetic field treatments improve germination and seedling growth in Festuca arundinacea Schreb. and Lolium perenne L. Seed Science and Technology, 36, 31–37.

Celestino, C., Picazo, M.L. & Toribo, M. (2000). Influence of chronic exposure to an electromagnetic field on germination and early growth of Quercus suber seeds; Preliminary study. Electromagnetobiology, 19, 115–120.

Dhawi, F. & Al-Khayari, J.M. (2009) Magnetic Fields Induce Changes in Photosynthetic Pigments Content in Date Palm (Phoenix dactylifera L.) Seedlings. The Open Agriculture Journal, 3, 1–5.

Faqenabi, F., Tajbakhsh, M., Bernooshi, I., Saber-Rezaii, M., Tahri, F., Parvizi, S., Izadkhah, M., Hasanzadeh Gorttapeh, A. & Sedqi, H. (2009). The effect of magnetic field on growth, development and yield of safflower and its comparison with other treatments. Research Journal of Agriculture and Biological Sciences, 4,174–178.

Florez, M., Carbonell, M.V. & Martinez, E. (2007). Exposure of maize seeds to stationary magnetic fields: Effects on germination and early growth. Environmental and Experimental Botany, 59, 68–75.

Foreman, J., Demidchik, V. & Bothwell, J.H. (2003). Reactive oxygen species produced by NADPH oxidase regulate plant cell growth. Nature, 27, 442–446.

Gidrol, X., Lin, W.S., Degousee, N., Yip, S.F. & Kush, A. (1994). Accumulation of reactive oxygen species and oxidation of cytokinin in germinating soybean seeds. European Journal of Biochemistry, 224, 21–28.

Gubbels, G.H. (1982). Seedling growth and yield response of flax, buckwheat, sunflower, and field pea after preseeding magnetic treatment. Canadian Journal of Plant Science, 62, 61–64.

Gurumoorthi, P., Senthil Kumar, S., Vadivel, V. & Janardhnan, K. (2003). Studies on agro botanical characters of different accessions of velvet bean collected from Western Ghats, South India. Tropical and Subtropical Agroecosystem, 2, 105–115.

Harichand, K.S., Narula, V., Raj, D. & Singh, G. (2002). Effect of magnetic fields on germination, vigour and seed yield of wheat. Seed Research, 30, 289–93.

Huang, H-H & Wang, S-R. (2007) The Effects of 60Hz Magnetic Fields on Plant Growth. Nature and Science, 5, 60–68.

Khan, M.M., Hendry, G.A.F., Atherton, N.M. & Vertucci-Walters, C.W. (1996). Free radical accumulation and lipid peroxidation in testas of rapidly aged soybean seeds: A light promoted process. Seed Science Research, 6, 101–107.

Kiranmai, V. (1994). Induction of mutations by magnetic field for the improvement of sunflower. Journal of Applied Physics, 75, 7181.

Lebedev, I.S., Litvinenko, L.G. & Shiyan, L.T. (1977). After-effect of a permanent magnetic field on photochemical activity of chloroplasts. Sov Plant Physiololgy, 24, 394–395.

Liszkay, A., Kenk, B., Schopfer, P. (2003). Evidence for the involvement of cell wall peroxidase in the generation of hydroxyl radicals mediating extension growth. Planta, 217, 658–667.

Liszkay, A., Van der Zalm, E. & Schopfer P. (2004). Production of reactive oxygen intermediates (O_2 -, H_2O_2, and ·OH) by maize roots and their role in wall loosening and elongation growth. Plant Physiology, 230, 3114–3123.

Martinez, E., Carbonell, M.V. & Amaya, J.M. (2000). A static magnetic field of 125 mT stimulates the initial growth stages of barley (Hordeum vulgare L.). Electromagnetic Biology and Medicine, 19, 271–277.

Martínez, E., Carbonell, M.V., Flórez, M., Amaya, J.M. & Ma-queda, R. (2009). Germination of tomato seeds under magnetic field. International Agrophysics, 23, 45–49.

Moon, J. & Chung, H. (2000). Acceleration of germination of tomato seeds by applying AC electric and magnetic fields. Journal of Electrostatics, 48, 103–114.

Muraji, M., Asai, T. & Wataru, T. (1998). Primary root growth rate of Zea mays seedlings grown in an alternating magnetic field of different frequencies. Bioelectrochemistry and Bioenergetics, 44, 271–273.

Phirke, P.S., Kubde, A.B. & Umbarkar, S.P. (1996). The influence of magnetic field on plant growth. Seed Science and Technology, 24, 375–392.

Phirke, P.S., Patil, M.R. & Tapre, A.B. (1990). Effect of magnetic field on germination of sunflower seeds. Agricultural Engineering Today, 14, 3–11.

Pietruzewski, S. (1993). Effect of magnetic seed treatment on yields of wheat. Seed Science and Technology, 21, 621–626.

Pittman, U.J. (1977). Effects of magnetic seed treatments on yields of barley, wheat, and oats in southern Alberta. Canadian Journal of Plant Science, 57, 37–45.

Podlesny, J., Pietruszewski, S. & Podlesna, A. (2004). *Efficiency of the magnetic treatment of broad bean seeds cultivated under experimental plot conditions. International Agrophysics, 18, 65–71.*

Podlesny, J., Pietruszewski, S. & Podlesna, A. (2005). *Influence of magnetic stimulation of seeds on the formation of morphological features and yielding of the pea. International Agrophysics, 19, 1–8.*

Radhakrishnan R. & Ranjitha Kumari B. (2013). *Influence of pulsed magnetic field on soybean (Glycine max L.) seed germination, seedling growth and soil microbial population. Indian Journal of Biochemistry and Biophysics. 50, 312–317.*

Rajendra, P., Nayak, H.S., Sashidhar, R.B., Subramanyam, C., Devendarnath, D., Gunasekaran, B., Aradhya, R.S.S. & Bhaskaran, A. (2005). *Effects of power frequency electromagnetic fields on growth of germinating Vicia faba L., the broad bean. Electromagnetic Biology and Medicine, 24, 39–54.*

Schroeder, J.I., Allen, G.J., Hugouvieux, V., Kwak, J.M. & Waner, D. (2001). *Guard cell signal transduction. Annual Review of Plant Physiology and Plant Molecular Biology, 52, 627–658.*

Shabrangi, A., Majd, A., Sheidai, M., Nabyouni, M. & Dorranian, D. (2010). *Comparing effects of extremely low frequency electromagnetic fields on the biomass weight of C3 and C4 plants in early vegetative growth. PIERS Proceedings, Cambridge, MA, July 5–8, 593–598.*

Shine, M.B. (2011). *Biophysical and physiological changes in maize (Zea mays L. var. HQPM1) after pre-treatment of seeds by magnetic field.* (Ph.D. thesis)

Shine, M.B. & Guruprasad K.N. (2012). *Impact of pre-sowing magnetic field exposure of seeds to stationary magnetic field on growth, reactive oxygen species and photosynthesis of maize under field conditions. Acta Physiologia Plantarum, 34, 255–265.*

Shine, M.B., Guruprasad, K.N. & Anjali, A. (2011a). *Enhancement of germination, growth and photosynthesis in soybean by pre-treatment of seeds with magnetic field. Bioelectromagnetics, 32, 474–484.*

Shine, M.B., Guruprasad, K.N. & Anjali A (2011b). *Superoxide radical production and performance index of Photosystem II in leaves from magnetoprimed soybean seeds. Plant Signalling and Behaviour, 6, 1635–1637.*

Shine, M.B., Guruprasad, K.N. & Anjali A (2012). *Effect of stationary magnetic field strengths of 150 and 200 mT on reactive oxygen species production in soybean. Bioelectromagnetics, 33, 428–437.*

Vashisth A, Nagarajan S (2007) Effect of presowing exposure to static magnetic field of maize (Zea mays L) seeds on germination and early growth characteristics. Pusa Agrisci 30: 48–55.

Vashisth A, Nagarajan S (2008) Exposure of seeds to static magnetic field enhances germination and early growth characteristics in chickpea (Cicer arietinum L.). Bioelectromagnetics 29:571–578

Vashisth A, Nagarajan S (2010) Effect on germination and early growth characteristics in sunflower (Helianthus annuus) seeds exposed to static magnetic field. J Plant Physiol 167:149–156

www.ingramcontent.com/pod-product-compliance
Lightning Source LLC
Chambersburg PA
CBHW061820210326
41599CB00034B/7057